Schistosomes

Schistosomes

Development, Reproduction, and Host Relations

PAUL F. BASCH

New York   Oxford
Oxford University Press
1991

Oxford University Press

Oxford   New York   Toronto
Delhi   Bombay   Calcutta   Madras   Karachi
Petaling Jaya   Singapore   Hong Kong   Tokyo
Nairobi   Dar es Salaam   Cape Town
Melbourne   Auckland

and associated companies in
Berlin   Ibadan

Copyright © 1991 by Oxford University Press, Inc.

Published by Oxford University Press, Inc.,
200 Madison Avenue, New York, New York 10016

Oxford is a registered trademark of Oxford University Press

Library of Congress Cataloging-in-Publication Data
Basch, Paul F., 1933–
Schistosomes: development, reproduction, and host relations / Paul F. Basch.
p. cm.   Includes bibliographical references and index.
ISBN 0-19-505807-0
1. Schistosomiasis.   2. Schistosoma.   I. Title.
[DNLM: 1. Host-Parasite Relations.   2. Reproduction.
3. Schistosoma—growth & development.
4. Schistosoma—parasitology. QX 355 B298s]
QR201.S35B37   1991   614.5′53—dc20   DNLM/DLC
for Library of Congress   91-3002

9 8 7 6 5 4 3 2 1

Printed in the United States of America
on acid-free paper

Preface

As parasites living within the circulatory systems of their vertebrate hosts, adult schistosomes face problems of reproduction and egg dispersal far beyond those of most other animals. Despite their many apparent frailties and the statistical improbability of survival through a daunting life cycle, schistosomes have developed mechanisms not only to survive, but to prevail, and have evolved to become significant pathogens of birds and mammals throughout the world.

From the point of view of humans, schistosomiasis is one of the six diseases selected as especially significant in the World Health Organization's "Special Programme for Research and Training in Tropical Diseases." The number of people infected with any of the three major or several minor forms of schistosomiasis is not known. Most authors accept the well-worn estimate of the World Health Organization of two hundred million cases, or about 4 to 5 percent of the world's population. Many more hundreds of millions live in endemic areas, at continuing risk of infection and disease.

The combination of biological uniqueness and biomedical immediacy has brought the schistosomes to the forefront of parasitology.

As with other fields of biology, parasitology stands today at a watershed, between the descriptive and functional biology of the past and the analytic and molecular biology of the future. It is my intention to facilitate the work of those investigators who wish to apply advanced techniques to schistosomes, but who lack familiarity with the fundamental biology of these organisms.

Not surprisingly, the accumulation of knowledge about schistosomes has been unplanned and uneven. Information is fragmentary in many areas as individual investigators, with primary interests in other fields of science, have dipped into the study of schistosomes to apply their specialized concepts and methods. The resultant literature, scattered and often uncoordinated, makes a comprehensive and critical analysis both frustratingly difficult and increasingly necessary. Students in particular, believing that the subject of their (or their mentor's) current research is of unique and eternal significance, may benefit from the biological and historical perspectives offered here. In the quest for novelty and currency, the fruits of previous investigations may well be overlooked, ignored, or needlessly repeated; and in the absence of a unifying overview today's innovations may encounter the same neglect tomorrow.

The treatment in this monograph is selective and interpretive rather than encyclopedic; e.g., no attempt was made to ferret out every published datum for the tables. My objective is to present the main features of schistosome biology, pointing out areas of ignorance, controversy, or special significance, and topics in need of elucidation. This book is grounded in an organismal viewpoint, with broad consideration of the schistosomes as distinctively adapted parasites. Because parasitism by definition involves interactions with hosts, it makes little sense to discuss schistosomes in abstract isolation. Yet a great proportion of published research on these organisms has focused on responses of marginally appropriate rodents as experimental proxies for humans. In the interest of brevity, I have decided to exclude most exclusively host-centered studies, such as:

case reports, community infection rates, and epidemiological studies, including mathematical modeling;

analyses of human behavior and water contact patterns;

control methods and organization of public health efforts;

methods and techniques for diagnosis and treatment of infection;

the effects of schistosome infection on nonadaptive host metabolic or pharmacologic responses;

the effects of host irradiation or other unnatural pretreatment on subsequent schistosome infections;

the intimate workings of host immunity, including the role of immunoglobulins, complement, cytokines, effector cells, etc.;

pathologic mechanisms *per se,* granuloma formation, and the passage of schistosome eggs through host tissue;

chemotherapeutic trials.

The bibliographies on schistosomiasis compiled by the World Health Organization (1960), Warren and Newill (1967), Warren (1973), Warren and Hoffman (1976) and Hoffman and Warren (1978) were extremely useful in searching the earlier literature. Although the list of references in this book is extensive, hundreds of other papers were read but, for various reasons, not used. I have not included most phenomenologic reports of the presence or absence of various substances, or applications of analytic techniques, where these do little to elucidate the fundamentals of schistosome biology.

Similarly, after long reflection I have decided not to include methods for rearing and infecting hosts in the laboratory, harvesting organisms, growing them in vitro, preparing them as specimens, and similar procedures, as these are virtually without end.

As this is not a taxonomic study of the schistosomes, scientific names are incorporated into the text as they were stated by the original author, without regard to later revision.

During the prolonged gestation of this monograph there appeared the multiauthor volume titled "The Biology of Schistosomes From Genes to Latrines" (D Rollinson and AJG Simpson, Editors), which reviews many of the subjects just mentioned and which serves as an excellent complement to the book now in your hands.

My gratitude goes to the Research Services staff at the National Library of Medicine in Bethesda, Maryland, who generously made available to me the unsurpassed facilities of that institution and provided patient and expert technical support. I thank my wife, Natalicia, and our sons, Richard and Daniel, for their indulgence during my lengthy periods of distraction. Rosario Villacorta helped mightily to cross-check everything and to maintain bibliographic integrity. Drs. Allen Cheever, Christopher Bayne, James McKerrow, and other colleagues, have read all or part of the manuscript, and have pointed out various omissions and questionable interpretations. As I have not always taken their advice, responsibility for those errors remaining in the text is mine alone. Finally, I owe cordial thanks to those scientific pioneers and colleagues whose inspiration and effort have wrested from nature the raw material for this still unfinished story.

Stanford, Calif.   P. F. B.
November 1990

Contents

Schistosomes

1

Overview

Schistosomes are trematodes, and possess the general characteristics of this large and diverse group of parasites. There is little question among zoologists that the ancestors of trematodes were free-living turbellarian or rhabdocoel flatworms, which became adapted as parasites of mollusks in very early times. The association between trematodes and mollusks is therefore long and intimate, having persisted over hundreds of millions of years. Archaegastropods, whose modern relatives include marine limpets, *Abalone*, and the like, are found in the earliest Paleozoic strata, and although we have no traces of fossil parasites, it is probable that by the beginning of the Paleozoic era, one-host trematode life cycles had already become well established. Those parasites may have resembled the Aspidogastrea, relatively primitive trematodes living today, which are often confined to a single molluscan host.

Higher Mesogastropods and Neogastropods radiated broadly in the Mesozoic: today these groups include many well-known marine snails such as conchs, whelks, and periwinkles. Without doubt their trematode parasites have accompanied these mollusks for millions of generations in an unending process of coevolution, and have diversified along with their hosts. Modern derivatives of those Mesozoic snails are now hosts to thousands of species of trematodes; of interest here are freshwater prosobranchs such as *Oncomelania*, the primary molluscan host of Asian schistosomes. Subsequently some gastropods invaded the terrestrial habitat, losing their operculum and gills as they evolved into the familiar land snails. Some of these became secondarily aquatic in fresh waters, but always remained as hosts for trematode parasites. Thus the air-breathing aquatic pulmonates such as *Biomphalaria, Bulinus, Lymnaea,* and other freshwater snails that are now hosts of African and New World schistosomes began their worldwide distribution.

Meanwhile, the vertebrates had begun to appear. Many species of fishes

3

Trematode Transmission Patterns

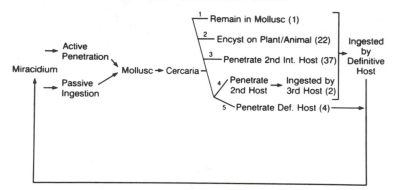

FIGURE 1-1 Five basic transmission patterns of digenetic trematodes as delineated by cercarial behavior. The numbers in parentheses represent the approximate number of families in which each pattern occurs. The schistosome pattern is number 5. *Source:* Shoop 1988.

were predators on aquatic mollusks, whose contained trematode parasites must have been ingested incidentally many millions of times. Gradually selection favored early proto-trematodes that were able to survive and reach sexual maturity within the fish gut and its branches, as the parasites became first resistant, then accustomed, then prosperous, and finally dependent on this habitat. Digenetic (two-host) life histories achieved dominance through their ability to exploit this food chain, utilizing their genetic plasticity to great advantage. Until today most adult trematodes live within the vertebrate intestine.

The coevolution of trematodes and their molluscan and vertebrate hosts (Wright and Southgate 1981) diversified the established patterns, gradually developing into the full range of two- and three-host trematode life histories described in detail below. A comprehensive summary of transmission patterns of extant trematodes was presented by Shoop (1988) (Fig. 1-1).

Yamaguti (1971) estimated the following numbers of trematode species living as adults in each class of vertebrates: Teleost fish, 2,800; Amphibia, 400; Reptiles, 750; Birds, 2,400; Mammals, 1,200.

Life Cycle

The term *life cycle* is universally applied to trematodes and other parasites, but it may be more accurate to think of a "life spiral." The world and its inhabitants never stop changing and evolving and each successive circuit through the trematode life history encounters, however imperceptibly, slightly different circumstances from that of the preceding generation. Each of the required two (or three) hosts, as well as the parasite itself, and the external environment, are continually undergoing minute alterations, which in the aggregate have led to the rich diversity found among the digenetic trematodes.

Modern trematodes have developed a great variety of methods for passage from one host to another, and for survival in each. Interpretations of the basis for the complex life cycles seem to be almost as numerous as the cycles themselves, but current thinking (e.g., Clark 1974) holds that the concepts of alternation of generations and polyembryony are unhelpful and that budding and metamorphosis are more useful to explain larval stages of digenetic trematodes.

Typically, adult flukes live within their vertebrate definitive host. A few live elsewhere, such as *Transversotrema* under the scales of fishes. Trematode eggs exit the host usually with feces, but sometimes in urine, sputum, or by other means. In some species the eggs, liberated into the environment, must be eaten by the snail first intermediate host. More commonly in aquatic species the eggs hatch after several days of incubation in the environment and release a swimming multicellular miracidium whose mission it is to locate and penetrate an appropriate mollusk in which it can develop. Within the tissues of this host (usually a snail, sometimes a bivalve), the miracidial body metamorphoses rapidly to a mother, or primary, sporocyst, which has limited motility and becomes saccular in form. From a germinal epithelium that lines the cavity of the sporocyst, balls of cells are budded off, which develop into a second intramolluscan larval stage. In some species these larvae are saccular and relatively passive daughter, or secondary, sporocysts; in others they develop into rediae, which possess a mouth, pharynx, and gut, and move actively within the molluscan host, eating bits of tissue and even preying actively on competing trematode larvae. One or more additional rounds of budding occur within these larvae, whether daughter sporocysts or rediae, giving rise eventually to tailed swimming larvae called cercariae, which escape from the mollusk and, depending on the species, play out a rigidly predetermined pattern of behavior. Some, such as the liver fluke *Fasciola,* attach to aquatic vegetation, whence they discard the tail and, while secreting an impermeable covering over the body, metamorphose slightly to become a metacercaria (metacercarial cyst). In a second pattern, some cercariae enter a specific second intermediate host such as (1) a snail of the same or different species, (2) an insect larva or other aquatic invertebrate, or (3) a fish. Here the metacercaria develops, active or encysted according to the species, and waits patiently for its transitory host to be eaten by an acceptable definitive host in which sexual maturity will ensue. A third pattern, followed by the schistosomes, has the cercaria penetrate directly through the skin and into the body of the definitive vertebrate host, bypassing a metacercarial stage. Still other, more complex, life histories may be found among the trematodes.

The plasticity of trematode reproductive patterns, with neoteny and truncation of developmental stages, is remarkable. For instance, miracidia are known that already contain daughter sporocysts or rediae before they have even entered their snail host. Some metacercariae become reproductively mature within an insect second intermediate host, completely cutting off the normal adult stage within a vertebrate (Macy and Basch 1972). More remarkably, the cercariae of *Proterometra dickermani* are not only sexually mature, but contain ova with active miracadia even before they have left their snail intermediate host (Anderson and Anderson 1963).

Structure and Function

The general biology of trematodes has been reviewed extensively by Erasmus (1972), and by Smyth and Halton (1983).

Digenetic trematodes share the general characteristics of the flatworms: they are bilaterally symmetrical, acoelomate (lack a body cavity), and are made up of a variety of cell types organized into complex organ systems.

When considering the structure of trematodes, we (as vertebrates) tend to think first of the familiar adult stage. However, the fertilized, or even unfertilized, ovum contains all the instructions necessary to construct and operate not only the adult but also all of the larval forms that (1) live and reproduce within the obligate mollusk host; (2) parasitize other intermediate hosts, if any; and (3) invade the external environment while passing between one host and the next. Each stage must face the challenges of functioning in its particular chemical, osmotic, and thermal environment. Microhabitats within the various hosts are marked by discrete physiological and immunological conditions, which may be hostile toward the parasitic invader. The trematode must be prepared to change from one stage to another almost instantly and with little warning. A virtuoso performance to say the least, and all the more remarkable in being accomplished with exuberant success by creatures at so low a position in the phylogenetic scale.

General Structure

The Digenetic trematodes are so variable in detail that one is at a loss to nominate a prototypical form. The typical adult trematode or fluke (Fig. 1–2) is dorsoventrally flattened and leafy, with an exposed epithelial outer tegument not covered by any impermeable superficial layer. This syncytial tegument, which lacks nuclei, is connected through cytoplasmic strands to cell bodies deep within the solid parenchyma of the animal (Fig. 1–3). The tegument may be marked by abundant sensory cells of several types, and by spines ranging from tiny and needle-like to relatively heavy and blunt, on a microscopic scale. The surface, which forms the obvious interface with the host, may serve both absorptive and secretory functions.

A mouth within the oral sucker at the anterior end leads through a muscular pharynx into paired intestinal ceca, which may be simple or branched. The ceca end blindly and as there is no anus the unassimilated remains of food must be regurgitated along with other gut contents. Depending on the species and location, adult trematodes feed on host intestinal contents, cells, secretions such as bile, or in the case of schistosomes, on blood.

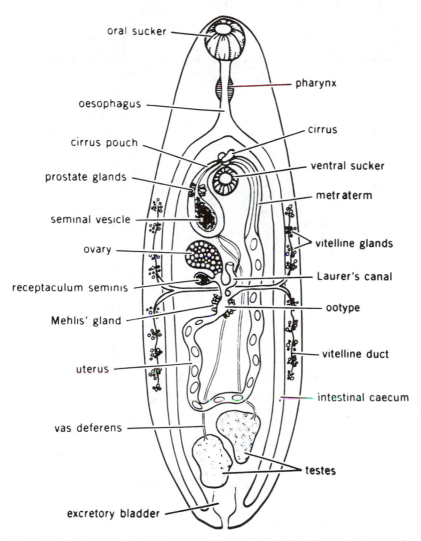

oral sucker

pharynx

oesophagus

cirrus

cirrus pouch

ventral sucker

prostate glands

metraterm

seminal vesicle

vitelline glands

ovary

Laurer's canal

receptaculum seminis

ootype

Mehlis' gland

vitelline duct

uterus

intestinal caecum

vas deferens

testes

excretory bladder

FIGURE 1–2 A generalized diagram of the anatomy of a digenetic trematode. *Source:* Smyth and Halton 1983.

A median ventral sucker or acetabulum is found on almost all adult trematodes, some distance posterior to the oral sucker. This muscular cup has no inner connection to the digestive system and functions exclusively as an attachment organ to maintain the position of the worm.

Numerous circular and longitudinal muscle fibers, found beneath the tegument and extending into or through the parenchyma, are responsible for the active but deliberate movements of many trematodes. All larval stages possess muscle fibers of the same types. Cercariae have well developed muscles in the body and tail by which they can move actively to reach the next host or encystment site.

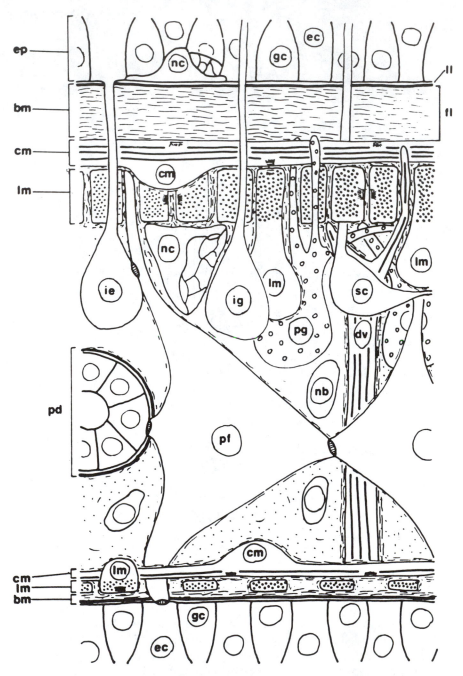

Figure 1–3. Model of configuration and common cell types as seen in many Platyhelminthes in the polyclad-neophoran line of evolution. Abbreviations: bm, basement membrane; cm, circular muscle cells; dv, dorsovental muscle cell; ec, regular epithelial cell; ep, epidermis; fl, fibrillar layer of basement membrane; gc, gland cells; ie, invaginated epidermal cells; ig, invaginated gland cells; ll, limiting layer of basement membrane; lm, longitudinal muscle cells; nb, neoblasts; nc, nerve cells; pc, peritoneal cells; pd, protonephridial duct; pf, fixed parenchyma cells; pg, pigment cells; sc, primary sensory cell. *Source:* Rieger 1985.

A central neuronal mass (Fig. 1–4) consists of two or more cerebral ganglia from which numerous unmyelinated sensory and motor nerve fibers extend throughout the body. The nervous system appears to operate on a familiar cholinergic basis, as acetylcholine and its associated enzymes have been identified in several species. Aminergic neurons have been found in several trematodes, involving serotonin, dopamine and other biogenic amines. The host-finding stages, miracidia and cercariae, possess prominent pitted or ciliated unicellular sensory structures. In the intramolluscan stages (sporocysts and rediae) these organs are far less developed.

A reproductive system is prominent in adult trematodes, almost all of which possess complete sets of both male and female gonads, ducts, and other structures. The female system contains accessory glands of several types that secrete both cellular and soluble components of the eggs. These structures may occupy a substantial portion of the total mass of the worm. Testes may be few or numerous, simple, or exuberantly divided or branched. The singular reproductive biology of schistosomes, in which sexes are separate, will be discussed in detail in subsequent chapters.

Osmoregulatory System. A system unique to certain lower invertebrates is the osmoregulatory or protonephridial system, sometimes called the "excretory" system. The basic element of this system is the flame cell (Fig. 1–5), a complex structure consisting of an end or cap cell, from which about 50 to 100 flagella project into a narrow cylindrical tubule. The synchronous beating of the flagella, viewed through an oil immersion lens, resembles the flickering of a flame. Flame cells are seen to best advantage in cercariae, but

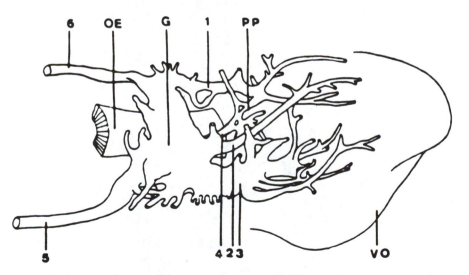

FIGURE 1–4 External view of the nervous system of adult male *S. mansoni;* reconstruction from serial sections by use of a camera lucida. G, ganglion; OE, esophagus; PP, periesophageal plexus; VO, oral sucker; 1, anterior dorsal nerve; 2, anterior lateral n; 3, anterior ventral n; 4, esophageal n; 5, posterior ventral n; 6 posterior dorsal n. *Source:* Dei-Cas et al. 1980.

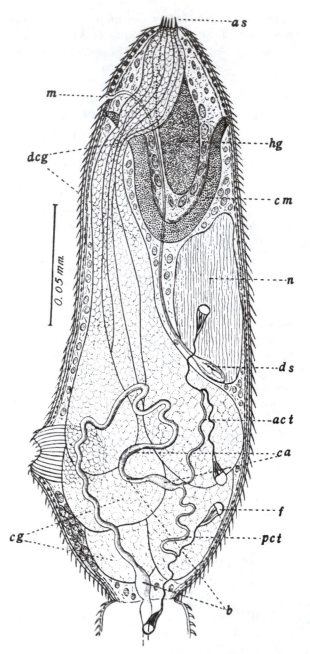

FIGURE 1–5 Cercaria of *S. japonicum*. Left, side view of body; Right, ventral view of entire cercaria. act, anterior collecting tube; as, anterior spines; b, excretory bladder; bt, excretory bladder of tail; ca, ciliated areas; cg, cephalic glands; cm, circular muscles; dcg, ducts of glands; ds, digestive system; exp, excretory pore; f, flame cell; g, penetration glands; hg, head gland; i, island in excretory bladder; m, mouth; n, nervous system; pct, posterior collecting tube; r, reproductive rudiment; s, sucker; st, stem of tail. *Source:* Cort 1921a.

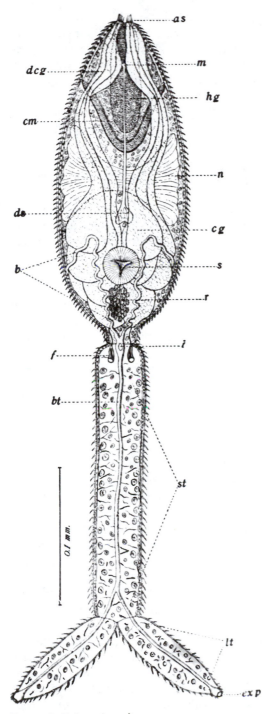

FIGURE 1–5 (*continued*)

occur in all stages. Tubules extending from each flame cell converge on each side into larger and larger ducts, which eventually lead into a median posterior dilatation or "excretory bladder," from which a single small duct opens onto the posterior tip of the worm. This system appears to function in removal of waste and in regulation of intercellular fluid composition.

In *Schistosoma mansoni* the parenchyma consists of (1) large nucleated cell bodies of overlying muscle fibers; (2) smaller muscle cells with cytoplasmic bodies; (3) cells connected to the nerve fibers, possibly neurosecretory; and (4) the bodies of tegumental cells (Reissig 1970). This cellular parenchyma is traversed by nerve axons, protonephridial tubules and ducts, branches of the intestine, and the various organs and conduits of the reproductive system.

Glands. In addition to the complex reproductive glands, unicellular secretory structures are found in the two transfer stages, associated with penetration of the hosts. The cercaria (Fig. 1–5) has at least three types of gland cells: cephalic, pre-acetabular, and post-acetabular, and some secretory function is found also in the tegumental cell bodies within the parenchyma (Cousin et al. 1981). The miracidium (Fig. 5–2) has a single apical gland cell and a pair of lateral gland cells (Pan 1980).

The ultrastructure of schistosomes has been described in numerous publications. The series of six papers by Silk and Spence (1969a, b), Silk et al. (1969), and Spence and Silk (1970, 1971a, b) although now old, still represent a key source of basic information.

Evolution of Schistosomes

It is commonly believed (Short 1983, Smith 1972) that schistosomes were derived from the hermaphroditic blood flukes of cold-blooded vertebrates, whose living representatives belong to the families Sangunicolidae of fishes and Spirorchiidae of reptiles (turtles). Azimov (1970) has reflected this view in a taxonomic revision that places all of the sanguinicolids, spirorchids, and schistosomes in a common order, the Schistosomatida, accepted by Smith (1972) in his comprehensive review of the blood flukes of poikilotherms. The blood flukes of fish and reptiles are hermaphroditic and inhabit the arteries, whereas the schistosomes, which live as adults only in mammals and birds, are dioecious (gonochoric) and, except for the aberrant *Dendritobilharzia*, live in the veins.

The present-day homeothermic vertebrates (mammals and birds) evolved from cold-blooded reptilian ancestors, and the schistosomes are very likely descended from spirorchid-like parasites of those ancient reptiles (Stunkard 1970, 1975). It seems highly probable that the transition from hermaphroditism to dioeciousness in blood flukes accompanied the acquisition of homeothermy by their hosts, perhaps dinosaurs or other reptiles transitional to birds and mammals, which radiated rapidly and broadly during the Mesozoic era, taking their blood flukes with them. Relatively rapid evolutionary changes in their definitive hosts forced the co-evolving parasites to adapt or become extinct, while at the same time remaining efficient parasites of their molluscan intermediate hosts (Basch 1990).

Phylogenetic relationships within the Schistosomatidae were considered in greatest detail by Carmichael (1984). Three major clades were revealed: one composed of the subfamilies Bilharziellinae and Gigantobilharziinae; a second consisting solely of *Macrobilharzia;* and a third including all of the other members of the Schistosomatinae. Birds were considered to be the primitive hosts of schistosomes, with at least two independent transfers to mammals.

A new subfamily, Griphobilharziinae, has been erected by Platt et al. (1991) to contain *Griphobilharzia* (Fig. 1–6), a new genus of schistosome from *Crocodylus johnstoni* in Australia. This parasite of freshwater crocodiles may represent a relic of an ancient schistosome line, or it may be a relatively recent adaptation to life in reptiles. Further analysis of this remarkable parasite should contribute significantly to an understanding of schistosome evolution.

Taxonomy of Schistosomes

Taxonomic keys and descriptions of schistosomes were presented by Witenberg and Lengy (1967), Skriabin (1951) (in Russian), and Yamaguti (1971), the latter two having numerous line drawings of the morphology of selected species. Farley (1971) still serves well as a review of prior classifications, and for descriptions and keys to genera and species except for the genus *Schistosoma* of mammals, for which Rollinson and Southgate (1987) provide a modern and detailed analysis. Two earlier comprehensive discussions of the general biology of schistosome complexes (Kuntz 1955, Malek 1961) are still worth reading.

All taxonomic parameters, from the family to the subspecies level, are matters of some controversy, and it is impossible to state the number of schistosome species with any reliability. Classical parasitologists have depended primarily on mensurable characters such as adult and cercarial morphology, egg size and shape (Fig. 1–7), and definitive and intermediate host range for guidance in delimiting species, but vagaries of adaptation and convergent evolution may confuse interpretations. For example, the egg of *S. sinensium* from Thailand and China (Fig. 1–8) greatly resembles that of *S. mansoni;* and primarily on that basis the two species were long considered to have close affinities. However, Greer et al. (1989) demonstrated by other characters that *S. sinensium* is actually more closely related to the *S. japonicum* group. The characterization of species differences at a biochemical (electrophoretic) level was carried out with *S. mekongi* and *S. japonicum* by Fletcher et al. (1980). These investigators concluded that the degree of biochemical difference between the two species is greater than that among various geographic isolates ("strains") of *S. japonicum*. Although the methods

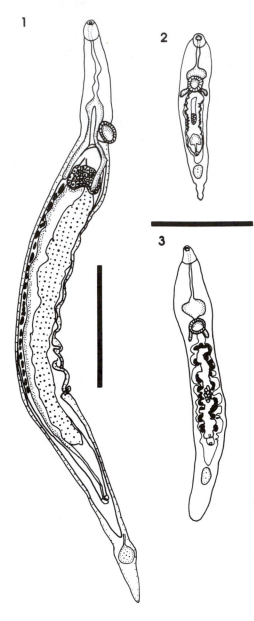

FIGURE 1–6 Adults of Griphobilharzia amoena from the freshwater crocodile Crocodylus johnstoni in Australia. 1, holotype, ventral view. Female is positioned laterally within the gynecophoric chamber. Large stippled area in the female is the uterus; 2, immature form, ventral view; 3, older immature form. Scale bar for Fig. 1 = 500 μm; for Figs. 2,3 = 200 μm. *Source:* Platt et al. 1991.

of molecular biology (McCutchan et al. 1984, LoVerde et al. 1985b, Walker et al. 1989b) can provide significant supporting information about the degree of relationship among schistosome populations, in the last analysis the definition and delimitation of species will depend on scholarly judgment.

A recent estimate of the number of different living kinds of schistosomes is that of Short (1983), who placed 83 species in "12 or more" genera in three subfamilies. More permissively, Yamaguti (1971) had listed approximately 103 species in 16 genera in 4 subfamilies. Tables 1–1 and 1–2, following Farley (1971), Short (1983) and Rollinson and Southgate (1987) suggest

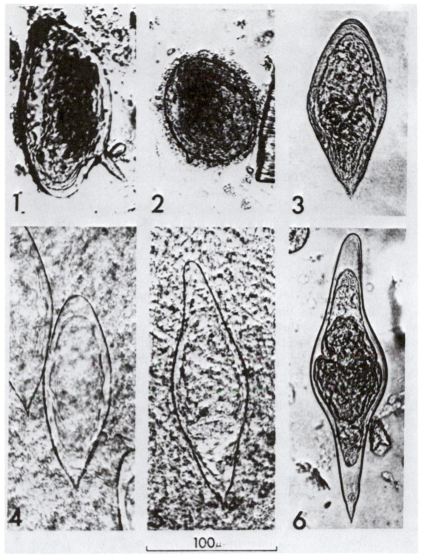

FIGURE 1–7 Eggs of 1, *S. mansoni;* 2, *S. japonicum;* 3, *S. haematobium;* 4. *S. intercalatum;* 5 *S. mattheei;* 6, *S. bovis. Source:* Jordan and Webbe 1969.

that there are approximately 85 different known species of schistosomes in mammals and birds. Additional species almost certainly await discovery.

Origin of Separate Sexes

It is difficult to conceive of an evolutionary continuum through which hermaphroditic ancestors have given rise to the dioecious schistosomes. Therefore it is not surprising that few authors have been willing to speculate on this subject. A karyological hypothesis has been advanced (Grossman et al.

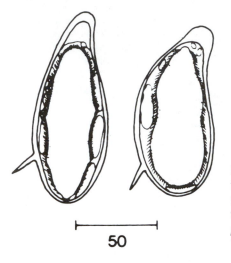

FIGURE 1–8 Egg of *Schistosoma sinensium* from northern Thailand. Note the superficial resemblance to that of *S. mansoni* even though this species is a member of the *S. japonicum* complex. *Source:* Greer et al. 1989.

50

1981a, Short 1983) in which it was assumed that primitive trematodes possess 10 or 11 pairs of telocentric or subtelocentric chromosomes with subsequent reduction in number by centromeric fusion. Chromosomal heteromorphism in spirorchids suggested differentiation of "pre-sex" chromosomes in species believed ancestral to schistosomes. Distinctly differentiated sex chromosomes have been reported in some schistosomes (Short 1983 for review), but here again nothing is known for the great majority of species of the family. It is likely that gender differentiation in chromosomal morphology accompanied, but did not induce, sexual differentiation in ancestral schistosomes.

Problems of Separate Sexes

Relative selective advantages must have outweighed the rigors inherent in gender separation, or else the schistosomes could not have undergone their vigorous evolutionary radiation. Difficulties to be overcome must have included the necessity for: (1) independent infection of the host by both male and female cercariae, which must find their way individually through the body to the site of a common rendezvous; (2) the necessity for males and females to remain physically associated for extended periods; (3) passage of the parasites, or at least of the females, to an advantageous oviposition site; (4) substantial reproductive wastage by eggs lost within host tissues; and (5) approximately equal success in infection of intermediate host snails by miracidia of both sexes.

Advantages of Separate Sexes. A strategic advantage of separate sexes is the mandatory cross-breeding and resulting heterosis arising from novel combinations of genetic factors. Schistosome evolution might be accelerated, for example, through acquisition of genetically diverse male and female cercariae by wide-ranging avian or mammalian hosts.

TABLE 1–1 Characteristics and Classification of the Schistosomes

Family Schistosomatidae

TREMATODA Dioecious, pharynx absent, esophagus short, paired ceca reuniting to form a common cecum extending to near posterior end of body. Suckers present or absent. Acetabulum when present anterior to genital pore in male. Body of male may be widened and infolded to form a gynecophoric canal. Testes of four or more follicles. Seminal vesicle present, with or without a cirrus pouch. Female with elongate ovary, sometimes spiral, lying anterior to the cecal union. Laurer's canal present or absent. Vitellaria extensive, extending from ovary to posterior end of body. Eggs nonoperculate, embryonated, with terminal or lateral spine or spineless. Parasitic in blood vessels of birds and mammals. About 85 species.

SUBFAMILY SCHISTOSOMATINAE Both suckers usually present. Male with well developed gynecophoric canal extending to posterior end of body. Testes anterior to cecal union. Genital pore immediately posterior to acetabulum in both sexes. Female slender with oval or spiral ovary, with or without a small oval seminal receptacle posterior to ovary. Parasitic in birds and mammals; about 39 species.

Genus *Schistosomatium* in rodents, North America. Intermediate host (IH), freshwater pulmonates (Lymnaeidae); 1 species.

Genus *Bivitellobilharzia* in elephants, India. Africa? IH, unknown; 2 species.

Genus *Heterobilharzia* in carnivores, rodents, lagomorphs North America. IH, freshwater pulmonates (Lymnaeidae); 1 species.

Genus *Austrobilharzia* in birds, cosmopolitan. IH, marine prosobranchs; about 6 species.

Genus *Orientobilharzia* in ungulates, equines, carnivores, camels, Near East to Asia. IH, freshwater pulmonates (Lymnaeidae); about 4 species.

Genus *Macrobilharzia* in birds, cosmopolitan. IH, unknown; about 2 species.

Genus *Ornithobilharzia* in birds. IH, marine prosobranchs or unknown; about 4 species.

Genus *Schistosoma* in mammals, cosmopolitan. IH, freshwater pulmonates (Planorbidae and Lymnaeidae); freshwater prosobranchs; about 19 species (see Table 1–2).

SUBFAMILY BILHARZIELLINAE Male and female similar, threadlike, or flattened. Gynecophoric canal absent or small. Testes numerous. Ovary very elongate in threadlike forms, compact in flattened species, in anterior half of body. Ootype with short uterus. Parasitic in birds; about 30 species.

Genus *Trichobilharzia*. In ducks or other birds, cosmopolitan. IH, freshwater pulmonates (Lymnaeidae or Physidae) or unknown; about 28 species.

Genus *Bilharziella*. In ducks (or other birds?) IH, freshwater pulmonates (Lymnaeidae or Planorbidae) or unknown; about 2 species.

SUBFAMILY GIGANTOBILHARZIINAE Male and female similar, threadlike, or flattened. Oral sucker absent or present, acetabulum always absent. Gynecophoric canal absent or small. Testes numerous. Ovary elongate in threadlike forms, compact in flattened species, in anterior half of body. Ootype with long uterus. Parasitic in birds; about 16 species.

Genus *Gigantobilharzia*. Cosmopolitan. IH, freshwater pulmonates, marine opisthobranchs, or unknown; about 12 species.

Genus *Dendritobilharzia* in ducks and pelicans. Cosmopolitan. IH, unknown; about 4 species.

SUBFAMILY GRIPHOBILHARZIINAE Male with large, enclosed gynecophoric canal forming a chamber, which begins slightly posterior to acetabulum and terminated anterior to a single testis, near posterior end of body. Female completely enclosed in gynecophoric chamber, oriented antiparallel to male.

Genus *Griphobilharzia*. From Northern Territory of Australia. In crocodiles. One known species.

Source: Adapted after Farley, 1971; Short, 1983; Rollinson and Southgate, 1987; Platt et al. 1991.

TABLE 1–2 Species of *Schistosoma*: Intermediate and Definitive Hosts

		Host	
Species	Distribution	Intermediate	Definitive
S. haematobium group			
*S. haematobium**	Af, adj	*Bulinus* (P)	Primates
	In	*Ferrissia* (A)	Primates
*S. intercalatum**	Af	*Bulinus* (P)	Primates
*S. mattheei**	Af	*Bulinus* (P)	Artiodactyla,
			Primates
S. bovis	Af, adj	*Bulinus,*	Artiodactyla
		Planorbarius (P)	
S. curassoni	Af	*Bulinus* (P)	Artiodactyla
S. margrebowiei	Af	*Bulinus* (P)	Artiodactyla
S. lieperi	Af	*Bulinus* (P)	Artiodactyla
S. mansoni group			
*S. mansoni**	Af, SA, Ca	*Biomphalaria* (P)	Primates
	Ca, Af, ME	*Biomphalaria* (P)	Rodents
S. rodhaini	Af	*Biomphalaria* (P)	Rodents, Carnivores
S. edwardiense	Af	*Biomphalaria* (P)	Artiodactyla
S. hippopotami	Af	unknown	Artiodactyla
S. indicum group			
S. indicum	In, SEA	*Indoplanorbis* (P)	Artiodactyla
S. spindale	In, SEA	*Indoplanorbis* (P),	Artiodactyla
		Lymnaea (L)	
S. nasale	In, SEA	*Indoplanorbis* (P)	Artiodactyla
S. incognitum	In, SEA	*Lymnaea, Radix* (L)	Rodents, Carnivores,
			Artiodactyla
S. japonicum group			
*S. japonicum**	C, J, Ph, T,	*Oncomelania* (O)	Primates,
	Is	*Oncomelania* (O)	Artiodactyla,
		Oncomelania (O)	Carnivores,
			Rodents,
			Perissodactyla
*S. mekongi**	SEA	*Tricula* (O)	Primates, Carnivores
*S. malayensis**	SEA	*Robertsiella* (O)	Primates, Rodents
S. sinensium	C, SEA	*Tricula* (O)	Rodents

Source: Adapted after Rollinson and Southgate, 1987.

Abbreviations: adj, adjacent areas; A, Ancylidae; Af, Africa; C, China; Ca, Caribbean; In, India; Is, Indonesia; J, Japan; L, Lymnaeidae; ME, Middle East; O, Prosobranchs; P, Planorbidae; Ph, Philippines; SA, South America; SEA, Southeast Asia; T, Taiwan.

*Found in humans

Although self-fertilization is possible in most digenetic trematodes, it seems that copulation between individuals is the rule whenever possible. Many groups of trematodes exhibit a strong tendency towards reciprocal mating between two hermaphroditic individuals. Inadequate reproduction in single-worm infections has been noted in philophthalmids (Fried 1962, Howell and Bearup 1980) and in *Paragonimus* (Sogandares-Bernal 1966), whose propensity for pairing is well known.

It may be argued that the mission of delivering their eggs within blood vessels may be accomplished equally well by hermaphroditic flukes, such as spirorchids and sanguinicolids; nevertheless, *res ipso loquitur:* the tactical advantage of genetically determined sexual separation is displayed as relative reproductive success. The female attains an augmented capability to place her eggs in a site in which the miracidial embryo develops safely, and from which they can exit the host to hatch quickly. The loss of eggs that are swept through the portal vein into the liver (at least in species such as *S. mansoni* and *S. japonicum*) is compensated for by avoiding the risk of predation and mishap faced by eggs that must undergo their entire development in the outside world. Meanwhile, in long-lived hosts, the female worm enjoys an extended and sheltered life.

A Possible Mechanism for Sexual Separation

The evolution of separate sexes may be associated with (1) the change in habitat from arteries to veins; (2) development of the homeothermal portal system, in which the smaller mesenteric venules may have different arborizations from those found in poikilothermal vertebrates; and (3) the adoption of primarily terrestrial hosts with fully developed lungs, rather than primarily aquatic hosts.

Selection toward separate sexes may have started by favoring adaptations that modified the timing of expression of the separate sets of genes controlling the male and the female reproductive systems, as well as genes for body musculature and locomotion. From a temporal viewpoint, it is likely that protandry, well known among trematodes (Fried 1986), and/or protogyny developed among hermaphroditic proto-schistosomes, resulting in individuals acting as functional males or functional females at different times. Selective processes can then have acted to suppress the expression of those genes controlling the development of the opposite gender-specific organ system, thereby establishing male-tending and female-tending genetic lines. Once established, each gender-predisposed genome evolved further morphological and functional refinements selected for by reproductive success, while inhibiting the contrasex genome. The degradation of the now nonfunctional genes for unused reproductive structures and processes of the repressed sex could readily have given rise to the separate males and females characteristic of modern schistosomes.

Although we will never know the sequence of intermediate stages undergone by ancestral schistosomes in the ancient past, the fact is that they were successful in establishing a unique pattern of bisexual reproduction within the Digenea. In the intervening eons, this basic pattern has become embellished by manifold adaptations dictated by the needs of particular life histories. A discussion of spousal functions in each sex, and of the dynamics of conjugal relations in modern schistosomes is found in Chapter 4. As an example, isolated females in the genus *Schistosomatium* are capable of producing parthenogenetic eggs (Short 1952a)—a clear benefit that reduces the need for male coinfection in short-lived rodent hosts (Loker 1983). One may

expect eventual selection of an evolutionary line in which uniparental reproduction is the rule and in which males are eliminated altogether, thus restor-
ing through a long and circuitous path a condition in a sense similar to that
found in the hermaphroditic ancestral flukes.

Chromosomes

The chromosomes of schistosomes (Fig. 1–9) have been reviewed and illustrated by Short (1983) in his presidential address to the American Society of
Parasitologists. Schistosomes are characterized by the ZZ/ZW chromosome
pattern, found also for instance in birds, in which the female is heteroga-
metic (Grossman et al. 1981a). As with other sexually reproducing organisms, the sex of offspring is determined in the zygote, with males generally
resulting from ZZ chromosomes and females from ZW. The basic diploid
number of chromosomes in schistosomes seems to be eight pairs with some

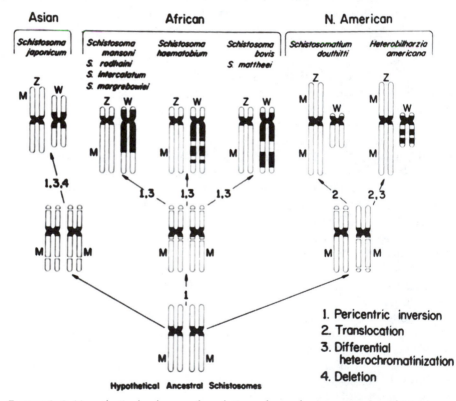

FIGURE 1–9 Hypothetical scheme of evolution of sex chromosomes in schistosomes.
The ideograms represent numbers of chromosomes, approximate centromere positions,
and show heterochromatic segments on W chromosomes. Numbers with arrows indicate hypothesized chromosomal mutations, listed lower right, during differentiation of
Z and W chromosomes. Source: Short 1983.

exceptions, such as *Schistosomatium,* which has seven pairs, and *Hetero-bilharzia,* which has ten. However, an isolate of *Heterobilharzia americana* has been found from Louisiana in which the male and female have different diploid chromosome numbers (male 20, female 19). This isolate has a ZZ male and ZWA female, whereas an isolate from adjacent Texas has 20 chromosomes in each sex and a conventional ZZ/ZW system (Short et al. 1987). Barshene et al. (1989) have performed karyotype analysis of *Schistosomatium, Austrobilharzia,* and two species of *Trichobilharzia* from northwestern Chukota (U.S.S.R.), finding the expected 14, 16, 16, and 16 chromosomes, respectively.

Radiation and Variety of Schistosomes

The classical representation of a pair of schistosomes in copula roughly resembles an anaconda in a canoe (Fig. 1–10). Although commonly held by biomedical parasitologists aware only of *S. mansoni, S. japonicum,* and *S. haematobium,* this paradigm represents but a small sampling from the broad range of schistosome variation. Indeed, in two of the three subfamilies (Bilharziellinae and Gigantobilharziinae) adult male and female worms are similar, both being either threadlike or flattened (Farley 1971). Regarding most described species, little is known except for their existence, but the schistosomes have clearly evolved an exuberant diversity in adult form while retaining conservatism in the pattern of life histories. Illustrations of many different schistosomes of birds and mammals are found in the works of Skriabin (1951) and Yamaguti (1958, 1971). Figure 1–11 illustrates some of the remarkable differences among the known species of schistosome parasites and argues eloquently for the need for additional studies on those species whose biology has been all but ignored.

The schistosome life cycle, already alluded to briefly, is, as far as is known, essentially the same for all species. Only two hosts are needed: de-

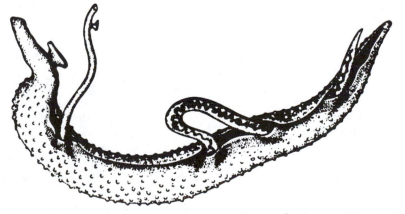

FIGURE 1–10 A pair of *Schistosoma mansoni* in copula. *Source:* Gönnert 1949a.

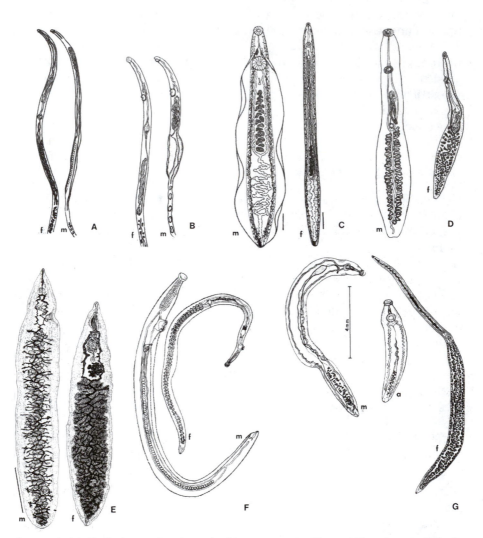

FIGURE 1–11 Radiation and variety of schistosomes. A, *Gigantobilharzia gyrauli,* both sexes long and attenuated. *Source:* Skriabin (1951) after Brackett; B, *Pseudobilharziella waubesensis,* male bearing small gynecophoral canal. *Source:* Skriabin (1951) after Brackett; C, *Austrobilharzia variglandis,* male with large gynecophoral canal. *Source:* Chu and Cuttress (1954). D, *Bilharziella polonica. Source:* Skriabin (1951) after Kowalewsky; E, *Dendritobilharzia pulverulenta,* both sexes are flattened and leaflike. *Source:* Ulmer and vandeVusse (1970). F, *Ornithobilharzia bomfordi,* male with numerous testes posteriorly located. *Source:* Skriabin (1951) after Montgomery. G, *Heterobilharzia americana,* with dimorphic adult males. Sickle-shaped male on left, attenuated adult male in center, female on right. *Source:* Lee (1962). a, attenuated male; f, female; m, male.

finitive (bird or mammal), and intermediate (snail). The definitive host passes eggs, usually via the feces but sometimes through urine (in *S. haematobium*) or the nasal secretions (in *S. nasalis*). In an appropriate aquatic environment these eggs hatch, often within minutes, and release swimming miracidia. The miracidia penetrate potential molluscan intermediate hosts, lose their ciliated epidermal plates within the snail tissues, and differentiate into mother sporocysts. One or more additional generations of daughter sporocysts is followed by production of swimming infective cercariae, which are released in large numbers by the snail host. All of the cercarial progeny of a single miracidium are of the same sex.

Once within the definitive host, the cercarial body gradually changes through a juvenile form known as a schistosomulum to the male or female adult. During this period lasting several weeks, the organism travels considerably in the host body, particularly in the liver and vascular system. Adults identify the appropriate site within the veins, and the female sets about producing and depositing eggs, which she does with clocklike regularity for the remainder of her days.

Significance of the Intermediate Host

Morphologic comparison of adult forms alone is insufficient for a comprehension of evolutionary relationships. At least of equal significance is knowledge of intermediate and definitive host range, intramolluscan stages, cercarial morphology, genetic and biochemical makeup, and other important analytic information. An example of the primitive state of current knowledge (or, better, ignorance), is *Trichobilharzia* of anatid birds, the largest genus in the schistosome family (Table 1–1), whose cercariae are responsible for widespread outbreaks of human dermatitis. For the majority of species not even the complete life cycle is known with certainty, and some species are described only from fragments of adult worms. According to Farley (1971), any attempt to key out this genus or even to suggest synonyms is futile "since with the data at hand this must rest upon minute structural differences in the adult forms."

The coevolution of digenetic trematodes and mollusks was considered at length by Wright (1971) and by Wright and Southgate (1981), who emphasized the species found in Africa. The significance of knowledge of intermediate hosts is illustrated by comparing the *S. japonicum* complex with other species in the genus *Schistosoma*. Although the adults of *S. japonicum* and its Asian cousins closely resemble those of the other mammalian *Schistosoma*, their snail intermediate hosts are radically different (Table 1–2). The *S. japonicum* complex utilize amphibious prosobranchs of the genus *Oncomelania*, and its relatives, whereas most other species within *Schistosoma* develop only in aquatic pulmonates. For example, *S. mansoni*, *S. rodhaini*, *S. haematobium*, *S. bovis*, *S. mattheei* and some others utilize snails of the family Planorbidae, and *S. spindale* larvae grow only in the closely related pulmonate family Lymnaeidae.

Because prosobranch and pulmonate snails are evolutionarily far apart, it seems unlikely that present day representatives of these two molluscan groups represent the separate continuous carriage of almost identical mammalian schistosome parasites from ancient geological times. Therefore the utilization by the *S. japonicum* complex of snail hosts different from the others is explainable through the following logical sequence:

1. Most known species of *Schistosoma* require pulmonate intermediate hosts in the families Planorbidae and Lymnaeidae, and it is likely that the modern genus has evolved within this large and widespread group of freshwater snails. The ancestors of the present-day *S. japonicum* complex, like most other *Schistosoma,* required pulmonate snail hosts.

2. In most freshwater habitats such as streams, ponds and marshes, pulmonate and prosobranch snails live side by side.

3. Trematode miracidia are genetically determined to develop only within a well-defined snail host, and miracidia that penetrate the wrong snail host are recognized and destroyed, as described in Chapter 5.

4. Nevertheless, miracidia commonly make this suicidal mistake, which is disastrous for that individual but advantageous for the species, because:

5. One day in the distant past a miracidium displayed a mutation in its surface configuration that happened to permit it to survive in a local prosobranch without being recognized, *and* it made the otherwise fatal error of entering such a snail, *and* because it was by chance preadaptated to the new host it escaped destruction, *and* it was successful in generating the usual intramolluscan stages. Had this particular modified miracidium entered its "normal" host, it would have been recognized as foreign, and annihilated.

6. The infected prosobranch snail released many cercariae, all of which carried the novel gene for altered intermediate host specificity. Although these cercariae were all of one sex, they penetrated into mammals and grew into adults that, through mating, distributed the gene for prosobranch specificity to miracidia of both sexes, some of which subsequently penetrated prosobranch hosts.

7. In this way a new prosobranch-specific evolutionary line was started, which with further local radiation has developed into the group of sibling species known today as the *S. japonicum* complex.

A similar process may have occurred in the isolated circumstance of human schistosomiasis, presumably owing to *S. haematobium,* in India, in which the host snail is a freshwater limpet of the family Ancylidae, genus *Ferrissia* (Gadgil 1963, Gadgil and Shah 1956, Hubendick 1958, Sathe et al. 1981, Bidinger and Crompton 1989).

The likelihood that potential snail hosts will become actual hosts by this process may be relatively greater in the schistosomes of migratory birds, which repeatedly distribute their parasite eggs in different kinds of water bodies spread over thousands of miles. Within the large schistosome genus *Gigantobilharzia,* which parasitizes many different birds, species are known that utilize marine prosobranchs, freshwater prosobranchs, or freshwater pulmonates of the families Physidae or Planorbidae. Some species of *Gigantobilharzia* indistinguishable as adults are separable only by their different snail intermediate hosts (Farley 1971).

Significance of the Definitive Host. The definitive vertebrate host range of most adult schistosomes seems less restricted physiologically than the intermediate snail host range. For example, Pesigan et al. (1958) found natural infections with *S. japonicum* in pigs, dogs, goats, cattle, and water buffaloes in addition to humans in the Philippines. Nevertheless, geographic isolates may differ in their epidemiologic relation to various vertebrate hosts. Whereas rodents are used routinely as hosts for *S. mansoni* in the laboratory, reports of natural infection of rodents are few—for example, Schwetz 1954 in the Congo, and Imbert-Establet 1982 on Guadeloupe.

Intraspecific Variations

Numerous authors have compared geographic "strains" of schistosomes for various traits, many of which are indicated on Table 1–3. Such comparisons must always be viewed with caution, for the characteristics listed may not be generalizable to all populations of that species from the same country or geographic area. Indeed, the terms *isolate* or *stock* should in most cases be used instead of "strain," which carries an unwarranted connotation of stability. As Fletcher et al. (1981c) have pointed out, geographic comparisons of traits should be based on several isolates per region. It is not surprising that laboratory stocks of *S. mansoni* isolated from different parts of a country the size of Brazil should differ in certain characters, as demonstrated for example by Valadares et al. (1980, 1981a), and Magalhães and de Carvalho (1973). However, consistent differences occur among natural isolates within the island of Puerto Rico (Fletcher et al. 1981a) or even the much smaller St. Lucia (Kassim et al. 1979). Indeed, certain constant differences are demonstrable among cercarial pools from different aquaria in the same laboratory (Fletcher et al. 1981b, Smith and Clegg 1979); from clones of cercariae derived from different unimiracidially infected snails (Cohen and Eveland 1984); or from individual worms (Fletcher et al. 1981b, Hsü and Hsü 1958a, McCutchan et al. 1984). The issue of biochemical strain variations in helminths in general, including schistosomes, was reviewed by Bryant and Flockhart (1986).

The validity of certain traits to distinguish between populations has been questioned; for example, Coles and Thurston (1970) considered that testes number was not a valid criterion in *S. mansoni.* Nevertheless, diversity is an inherent characteristic that enhances the probability of survival of schis-

TABLE 1–3 Intraspecific Differences in Schistosome Populations

Species/Origin	Number of Isolates	Host	Variant Traits	Reference Note
S. mansoni				
Puerto Rico, Brazil, St. Lucia, Tanzania	5	Mouse	a, b, d	1
Africa, W. Hemisphere		Various	p	2
Puerto Rico, Liberia	2	Mouse	e, k	3
Liberia, W. Hemisphere	2	Guinea pig	g	4
Egypt, Sudan, Kenya, Uganda, S. Africa, Liberia, Puerto Rico, St. Lucia		Mouse	o	5
Puerto Rico	3	Mouse	o	6
E. & S. Africa, S.W. Asia, S. America, W. Indies	22	Mouse, Hamster	o	7
St. Lucia	2	Mouse	b, d	8
Brazil	2	Mouse	f, k	9
Puerto Rico, Tanzania	2	Monkey	a, d	10
Puerto Rico, Tanzania, Egypt	3	Mouse, Hamster	a, d, g, l	11
Puerto Rico, Tanzania, Egypt	3	Mouse	f	12
Puerto Rico, Tanzania, Egypt	4	Mouse, Hamster	a, b, d, h, l	13
Brazil	2	Mouse	d	14
Puerto Rico, Brazil, Egypt, Tanzania	4	Mouse	d, h, i, m	15
S. mansoni				
Guadeloupe, Brazil	2	Snail	u	16
Puerto Rico, Liberia	2	Mouse	t	17
Puerto Rico	2	Mouse	o	18
S. haematobium				
W. Africa, Sudan, Rhodesia		Various	c	19
Egypt, Tanzania	2	Hamster	a, b, d, e, i, j, k	20
Ghana	7	Mouse	j, p	21
Nigeria, Iran, Egypt, Tanzania	4	Hamster, Baboon	p	22
Nigeria, Iran	2	Hamster, Baboon	b, d, e, h, i, k	23
Iran, Mauritius, Ghana	4	Hamster	a, b, d, e, h, i	24
Africa, S.W. Asia	22	Hamster	o	25
Nyombe, Cairo	2	Hamster?	o	26
S. bovis				
Sardinia, Kenya	2	Mouse	g	27
Spain, Morocco, Sardinia, Iran, Kenya	10		o	28

TABLE 1–3 (continued)

Species/ Origin	Number of Isolates	Host	Variant Traits	Reference Note
S. leiperi				
Zambia, Botswana	2	Hamster	o	29
S. intercalatum				
Lower Guinea, Congo	2	Hamster	p	30
Cameroon, Zaire	2	Mouse	d, f, h, i	31
S. japonicum				
Japan, Philippines, Taiwan	3	Mouse	d	32
Japan, Philippines, Taiwan, China	4	13 species	f,g	33
	4	Mouse	q	34
	4	4 species	c	35
	4	Mouse	a	36
	4	Mouse	h,s	37
	4	Mouse	l	38
	4	Mouse	d,l	39
	3	Mouse	o	40
S. japonicum and S. mekongi				
Japan, Philippines, Taiwan, China, Laos	5	Mouse	c	41

Variant traits: a, prepatent period; b, fecundity; c, egg shape and/or dimensions; d, egg deposition patterns in host organs; e, uterine egg count; f, testes number and distribution; g, tendency to hermaphroditism in males; h, worm return (proportion of cercariae developing to adults); i, growth rate; j, adult worm sex ratio; k, adult worm dimensions; l, pathogenicity to host; m, egg granuloma size in liver; n, reciprocal immunity; o, isoenzyme patterns; p, snail host specificity; q, hybridization and infertility; r, survival in mammalian host; s, mammalian host specificity; t, drug susceptibility; u, cercarial shedding pattern.

Reference notes:1, Anderson and Cheever 1972; 2, Basch 1976; 3, Bruckner and Schiller 1974; 4, Buttner 1950; 5, de Boissezon and Jelnes 1982; 6, Fletcher et al 1981a; 7, Fletcher et al. 1981b; 8, Kassim et al. 1979; 9, Magalhães and de Carvalho 1973; 10, Nelson and Saoud 1968; 11, Saoud 1965; 12, Saoud 1966a; 13, Saoud 1966c; 14, Valadares et al. 1980, 1981a; 15, Warren 1967; 16, Théron and Combes 1983; 17, Lee 1972; 18, LoVerde et al. 1985a; 19, Alves 1949; 20, Paperna 1968; 21, Paperna 1968; 22, Webbe and James 1971a; 23, Webbe and James 1971b; 24, Wright and Knowles 1972; 25, Wright and Ross 1983; 26, Wright and Southgate 1976; 27, Saoud 1964; 28, Ross et al. 1978; 29, Southgate et al. 1981; 30, Wright et al. 1972; 31, Bjorneboe and Frandsen 1979; 32, Cheever and Duvall 1982; 33, Hsü and Hsü 1957; 34, Hsü et al. 1962; 35, Hsü and Hsü 1958a; 36, Hsü and Hsü 1958b, Hsü et al. 1960; 37, Hsü and Hsü 1960; 38, Hsü and Hsü 1960; 39, Warren and Berry 1972; 40, Ruff et al. 73; 41, Voge et al. 78a.

tosomes. Species must therefore not be viewed as monolithic, but as cumulations of heterogeneous local populations (Basch 1975) from which stocks are drawn by chance into the laboratory.

The situation is most complex in the case of those schistosomes of Africa and southwest Asia having terminally-spined eggs and bulinid intermediate hosts (such as *S. haematobium, S. bovis, S. intercalatum, S. mattheei*). Here species definition is often unclear, and there are many local forms, some of which may be interfertile, distinguishable primarily by size and shape of the eggs (e.g., Alves 1949, Wright and Southgate 1976, Kruger, Schutte et al. 1986), or by morphology of the male in *S. mattheei* (Kruger, Hamilton-Atwell, and Schutte 1986).

Table 1–3 is intended more as an indication of the traits observed to be variable than as an exhaustive compilation of specific data. Although the primary axis of classification is geographic origin, there are many other possible determinants of the differences displayed. Some of these confounding determinants are:

the history of the stock since isolation: mammalian and molluscan hosts and number of laboratory passages;
number and sex ratio of infecting cercariae;
number and age of adult worms recovered;
genetic characters, age, sex (?), and condition of hosts;
genetic variability within the schistosome population in the laboratory.

Infraspecific Variation in Chromosomes

Little work has been done on regional chromosomal variations within the same species. Short et al. (1989) looked at seven isolates of *S. mansoni* from the Caribbean, South America, and Africa, using conventional Giemsa staining and C-banding techniques. They found a constant number (16) with only minor differences in relative lengths and centromeric indexes.

Species "Crosses"

Because schistosomes are dioecious, hosts may be infected with males of one species and females of another. Alternatively, hosts may become infected with both sexes of two or more species, establishing the conditions for unidirectional or reciprocal outcrossings as well as conspecific matings. The previously mentioned complex of schistosomes having terminally spined eggs (Pitchford 1965) provides abundant opportunity for natural interbreeding in mammalian hosts including man (Wright et al. 1974). Such hybridization must be distinguished from "accidental" infection of humans with zoophilic schistosomes such as *S. bovis* and *S. mattheei* that normally infect livestock (Blair 1966a).

Definition of species in schistosomes is a subtle art. Even though specific markers may be identified via biochemical, immunological, or molecular techniques, taxonomic clarity may be unattainable in species complexes undergoing active evolution. Therefore there may always be some doubt as to the biological frontiers that separate one full species from another.

The concept of hybridization is particularly complicated in the schistosomes because of the stimulatory effect even of transgeneric pairing upon the female (Basch and Basch 1984) and the consequent possible confusion of parthenogenesis, apomixis, or pseudogamy with true genetic fertility or interfertility. Moreover, reciprocal matings are often disparate (Wright and Southgate 1976). According to Taylor et al. (1969), it is probable that parthenogenesis is the normal consequence of cross pairing in mixed infection and that hybridization rarely occurs.

Egg morphology has often been considered as a criterion of hybridization, but as Wright and Ross (1980) have pointed out, "the shape of the eggs produced is not necessarily a guide to the genetic constitution of the enclosed larvae." The egg shell is shaped by the ootype of the female; thus the shape of the egg shell from an initial cross will be determined by the genetic constitution of the mother (= matroclonal) even though the contained embryos may be hybrids. In such cases the developing embryo is of the F1 generation, but the egg shell, or at least its shape, is of the P1 generation. Wright and Southgate (1976) have clearly illuminated this point and shown how it had led to much confusion in the earlier literature.

Natural hybridization between species of *Schistosoma* has been reviewed by Southgate and Rollinson (1987) for the following combinations: *S. haematobium* × *S. matheei*; *S. haematobium* × *S. intercalatum*; *S. curassoni* × *S. bovis*. These authors have also reviewed laboratory-derived interspecific hybrids. Viable reciprocal hybrids between *S. bovis* (Kenya) and *S. curassoni* (Senegal) were produced by Rollinson et al. (1990). Confirmation of actual hybridization (not simply companionate induction of parthenogenesis in the female) was made by demonstration of intermediate electrophoretic enzyme patterns. Tables 1–4 and 1–5 summarize published data on natural and experimental schistosome species crosses with evidence of hybridization wherever found.

Molecular Biology

There is little doubt that molecular biology will be a dominant area of schistosome research in the coming years, and that many aspects of schistosome biology will be elucidated through molecular genetic analysis. Indeed, it is a primary goal of this monograph to provide a sound underpinning of biological information for those molecular biologists who have become interested in the schistosomes but lack an intimate familiarity with this group of parasites.

Because of (1) the rapid advancement of this field; (2) the frequent appearance of reviews; and (3) your author's limitations, it is impossible to provide more than a glimpse of schistosome molecular biology in this volume. The essentials of this subject were summarized by Simpson (1987). More recently, LoVerde et al. (1989) have reviewed developmentally regulated gene expression in *Schistosoma*. The single genome of each organism must function appropriately and almost instantaneously in each of the grossly disparate environments in which the various life history stages find themselves. LoVerde et al. used subtractive hybridization to create a probe, pSMf 61–46, whose corresponding mRNA occurs only in paired mature females, not in unisexual females, in males, or in eggs, and which is related to production of eggshell protein. A second clone, F4, codes for a different but related eggshell component.

TABLE 1–4 Experimental Coinfections of Schistosome Species

Species		Host	Result	Evidence	Reference Note
Male	Female				
douthitti	*mansoni*	Mouse	P, m, E, H		1
americana	*mansoni*	Mouse	P		
douthitti	*mansoni*	Mouse	P		
americana	*mansoni*	Mouse	P		
douthitti	*mansoni*	Mouse	P, m, E, H; F3		2
mattheei	*mansoni*	?	P		3
mansoni	*rodhaini*	Mouse	P, E, H, F		4
rodhaini	*mansoni*	Mouse	P, E, H, F	3	
haematobium	*intercalatum*	Hamster	P, M, E, H, F7	2, 3	5
douthitti	*mansoni*	*Peromyscus*	P,,M, E		6
mansoni	*douthitti*	*Peromyscus*	P (E)		
haematobium	*intercalatum*	Hamster	P	1	7
intercalatum	*haematobium*	Hamster	P	1	
haematobium	*intercalatum*	Hamster	P, M, E, H, F	2	8
intercalatum	*haematobium*	Hamster	P, M, E, h		
mattheei	*mansoni*	Mouse, Hamster	P		9
mansoni	*mattheei*	Mouse, Hamster	P, M, E, H; F3		
mansoni	*bovis*	*Mastomys*	P, M, E, H		
mattheei	*bovis*	Hamster	P, M, E, H; F2	2,3	
bovis	*mattheei*	Hamster	P, M, E, h		
haematobium	*mattheei*	Hamster	P, M, E, H; F3	2,3	
haematobium	*bovis*	Hamster	P, M, E, H; F3	2,3	
rodhaini	*mansoni*	Mouse, Hamster	P, M, E, H; F3	2,3,5	
mansoni	*rodhaini*	Mouse, Hamster	P, M, E, H; F3	2,3,5	
mansoni	*mattheei*	Mouse	P, M, E, h		10
japonicum	*mansoni*	Mouse, Hamster	P, M, E		11
mansoni	*japonicum*	Mouse, Hamster	P, M, E		
haematobium	*mansoni*	*Rhesus*	P, M, E		
mansoni	*haematobium*	Mouse	P, M, E		
japonicum	*haematobium*	Mouse	P, M, E		
japonicum	*mansoni*	Mouse, Hamster	P, M, E, h; F2		12
mansoni	*japonicum*	Mouse, Hamster	P, M, e		
intercalatum	*haematobium*	Hamster	P, M, E, H	1,2,3,4	13
haematobium	*intercalatum*	Hamster	P, M, E, H	1,2,3,4	
haematobium	*mattheei*	Hamster	P, M, E, H	1,2,3	14

TABLE 1—4 (continued)

| Species | | | | | |
Male	Female	Host	Result	Evidence	Reference Note
mansoni	*rodhaini*	Mouse	P, M, E, H	6	15
bovis	*curassoni*	Sheep, Mouse	P, M, E, H, F4	1,5	16

Species: *americana* = *Heterobilharzia americana*; *douthitti* = *Schistosomatium douthitti*; others are species of *Schistosoma*.

Result: E, egg production; Fn, filial generations achieved; H, hatching of miracidia; M, maturation; P, pairing. Lower case letter indicates weak or poor result.

Evidence: 1, isoenzymes; 2, egg characters; 3, snail host specificity; 4, mammalian host specificity; 5, adult worm characters; 6, cercarial emergence rhythms.

Reference notes: 1, Armstrong 1965; 2, Basch and Basch 1984; 3, LeRoux 1949; 4, LeRoux 1954; 5, Mutani et al. 1985; 6, Short 1948b; 7, Southgate et al. 1982; 8, Southgate et al. 1976; 9, Taylor 1970. 10, Taylor et al. 1969; 11, Vogel 1941a; 12, Vogel 1942a; 13, Wright and Southgate 1976; 14, Wright and Ross 1980; 15, Théron 1989; 16, Rollinson et al. 1990.

TABLE 1—5 Natural Hybridization of Schistosome Species

Species	Locality	Host	Evidence	Reference Note
S. haematobium plus				
S. mattheei	Hypothetical	H	b	1
	Possible, Zimbabwe	H	b	2
	Transvaal	H	b, d	3
	Transvaal	H	a, b	4
	Transvaal	H	b	5
	S. Africa, various	H	b	6
	S. Africa, various	H	e	7
S. intercalatum	Cameroon	H	b, c	8
	Cameroon	H	c	
	Gabon	H	b	
S. mattheei or *S. bovis*	Transvaal	H	b	9
S. bovis plus *S. curassoni*	Senegal	S	f	10
S. bovis plus *S. curassoni*	Senegal, Mali	C	f	11

Evidence: a, isoenzymes; b, egg shape and dimensions; c, snail host infectivity; d, mammalian host infectivity; e, presence or absence of tubercle on male; f, isoenzyme patterns.

Host: C, cattle; H, human; S, sheep.

Reference notes: 1, Alves 1947; 2, LeRoux 1954; 3, Pitchford 1961; 4, Wright and Ross 1980; 5, Kruger et al. 1986a, 86b; 6, Southgate et al. 1976; 7, Wright et al. 1974; 8, Burchard and Kern 1985; 9, Pitchford 1959; 10, Southgate et al. 1985; 11, Rollinson et al. 1990. See also Southgate and Rollinson 1987.

Simpson et al. (1982) isolated and characterized the DNA of *S. mansoni*. Adult and cercarial DNA were indistinguishable; the haploid genome contained 2.7×10^8 base pairs, about 10% of that found in mammals. Except for a failure to detect modified bases, the schistosome genome was similar to that of other eukaryotic species. There were several highly repetitive sequences. Genes coding for ribosomal RNA were repeated 500 to 1000 times. Using cloned segments of these ribosomal genes, McCutchan et al. (1984) developed radioactive probes that were utilized through a Southern blotting technique for comparison of species, "strain," and sex-specific genetic markers. Restriction analyses of DNA from *S. mansoni* from eight different geographic areas showed variations in copy number or minor fragments containing the ribosomal gene. Detection of these variations was used both for strain differentiation and for analysis of differences among individual worms. Sex-linked markers were also detected. In addition, the DNAs of *S. haematobium* and *S. japonicum* could be distinguished by differences in the length of the major repeating units and in array of minor bands. Ribosomal RNA gene probes were used to differentiate *S. haematobium* from related species by Walker et al. (1986), and Rollinson et al. (1986), in the same laboratory, discussed general applications of recombinant DNA techniques to problems of helminth identification.

The molecular and genetic basis of reproduction is an exciting and expanding area of schistosome biology. Developmentally regulated gene expression in female *S. mansoni* has been studied in several laboratories, with resultant reports of specific mRNA related to egg or eggshell production, found solely or primarily in sexually mature paired females. Several such genes have been identified, sequenced, cloned, and expressed (Blanton et al. 1985; Bobek et al. 1986, 1989; Johnson et al. 1987; Knight et al. 1986; Kunz and Symmons 1987; Meusel et al. 1987; Nene et al. 1986; Reis et al. 1989; Rumjanek 1989; Rumjanek et al. 1987; Simpson and Knight 1986; Simpson et al. 1987). These reports represent early steps in unraveling the complex reproductive biology of schistosomes.

Other analyses on a molecular genetic level might focus on the following interrelated topics: (1) the differences in reproductively related gene expression cascades between male-dependent females, such as those of *Schistosoma mansoni* and male-independent females, as in *Schistosomatium douthitti;* (2) the complex process of maturation (Nara et al. 1990), oogenesis, egg production, and deposition in a search for specifically destructive antischistosomal agents; (3) the genetics of drug resistance (Bruce et al. 1987; Brindley et al. 1989; Dias and Olivier 1985; Dias et al. 1988; Kinoti 1987); (4) the evolutionary mechanism of sexual separation from a hermaphroditic ancestor, including an estimation of the extent of unexpressed opposite-sex genome carried by males and females of different schistosome species and isolates; (5) an analysis of the developmentally regulated expression of partial hermaphroditism, as is found in certain isolates of *S. mansoni* in guinea pigs; and (6) the entire host-parasite interplay, including the influence of host hormones in gene expression. In addition, the identification of schistosomal genes coding for protective antigens, and their subsequent cloning in appropriate vectors, is a significant current goal of many laboratories engaged in the search for an antischistosomal vaccine (e.g., el-Sherbeini et al. 1990, Henkle et al. 1990, Loison et al. 1989).

2

Relations with Vertebrate Hosts

The Cercaria

Structure

Cercarial structure and cell biology have been reviewed recently by Basch and Samuelson (1990). Free swimming cercariae (Fig. 1–5) consist of a body about 125 μm long, which coverts eventually into the adult schistosome, and a forked tail about 200 μm long, which is discarded at the moment of penetration of the body into the vertebrate host. Both body and tail are covered by a single continuous syncytial tegument about 0.5 μm thick over the cercarial body and 0.2 μm on the tail, and characterized by a trilaminate plasma membrane over a trilaminate basement membrane. The nuclei, ribosomes, and other organelles of the tegument are located in subtegumentary extensions of the surface syncytial cells. The entire tegument (Fig. 2–1), in turn, is covered by a dense, fibrillar carbohydrate-containing glycocalyx about 1 to 2 μm thick (Figs. 2–2, 2–3) which may protect the cercaria from the low osmotic pressure of fresh water as it journeys between the molluscan and vertebrate internal environments.

The cercaria contains a neural mass (Fig. 2–4) and well-developed surface sensory structures, a flame-cell (osmoregulatory) system, and nonfunctional anlagen of digestive and reproductive systems from which the corresponding adult organs are to develop. Preacetabular, postacetabular, and head glands provide secretions necessary for cercarial functioning and host penetration. These transitory unicellular glands disappear after discharge and play no role in subsequent development. The bodies of male and female cercariae are morphologically indistinguishable. However, cell nuclei are shown so clearly in the cercarial tail that the sex, at least in *Schistosoma mansoni,* may be distinguished microscopically by a simple C-banding pro-

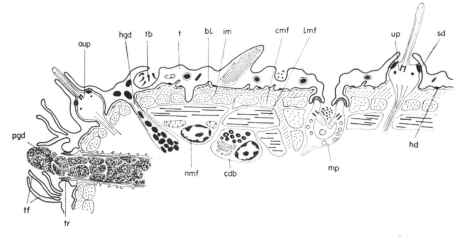

FIGURE 2–1 Relationship between the schistosome cercarial tegument and its associated structures. aup, anterior sensory organelle: unciliated papilla with short, ensheathed cilium; bl, basal lamina; cdb, cell containing dense bodies; cmf, circularly arranged muscle fibers; hd, hemidesmosome; hgd, head gland duct; im, interstitial material; lmf, longitudinally arranged muscle fiber; mp, multiciliated pit sensory organelle; nmf, nucleus of muscle fiber; pgd, penetration gland duct; sd, septate desmosome; t, tegument; tb, tadpole-shaped bodies in tegumentary vacuole adjacent to head gland; tf, tegumentary folds; tr, tegumentary rim round penetration gland aperture; up, unciliated papilla sensory organelle. *Source:* Hockley 1973.

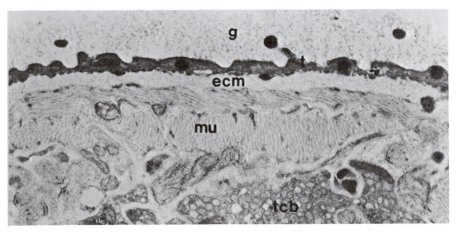

FIGURE 2–2 Surface of the cercaria of *S. mansoni.* Transmission electron photomicrograph, × 21,000. The glycocalyx (g) appears as a 1–2-μm thick fibrillar mesh. ecm, extracellular matrix; mu, muscle; tcb, tegumental cell bodies filled with membranous vesicles. *Source:* Samuelson and Caulfield 1985. Reproduced from the Journal of Cell Biology, 1985, Vol. 100, pages 1423–1434 by copyright permission of the Rockefeller University Press.

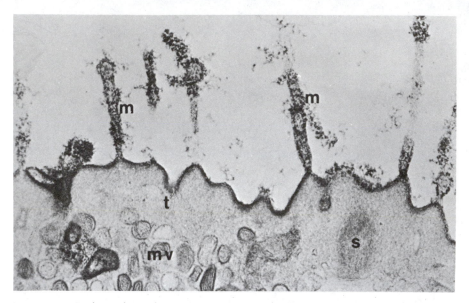

FIGURE 2–3 Surface of transforming cercarial body of *S. mansoni*. Transmission electron photomicrograph, × 50,000. The surface at 30 minutes is covered with microvilli (m). Within the tegument (t) are spines (s) and multilaminate vesicles (mv). *Source:* Samuelson and Caulfield 1985. Reproduced from the Journal of Cell Biology, 1985, Vol. 100, pages 1423–1434 by copyright permission of the Rockefeller University Press.

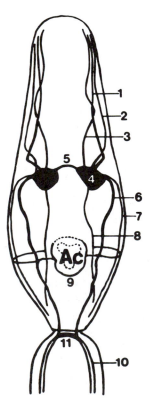

FIGURE 2–4 Schematic pattern of the nervous system in schistosome cercariae revealed by glyoxilic acid induced fluorescence. Ac, acetabulum; 1,2,3, antero-ventral, anterolateral, and antero-dorsal longitudinal nerve cords; 4, cerebral ganglion; 5, cerebral transverse commissure; 6,7,8, postero-ventral, postero-lateral, and postero-dorsal longitudinal nerve cords; 9; posterior transverse commissure; 10, tail lateral longitudinal nerve cord; 11, tail transverse commissure. *Source:* Orido 1989.

cedure that stains the extra volume of heterochromatin characteristic of the female nucleus (Liberatos and Short 1983). More complex molecular methods have also been described to accomplish the same end (Walker et al. 1989a, Webster et al. 1989).

Behavior. Cercariae of schistosomes swim with a rapid vibratile motion. Those of some species, such as the *S. mansoni* and *S. haematobium* groups, remain within the water volume, swimming both forwards and backwards. Others, including *S. japonicum, Schistosomatium douthitti,* and *Trichobilharzia ocellata,* display negatively geotactic and positively phototactive behavior and habitually rise to the surface where they hang suspended for long periods in the surface film. Such behavior both exposes cercariae to potential hosts and saves a substantial amount of energy that otherwise would be used in swimming. Cercariae of many schistosomes, such as *Trichobiharzia,* are termed "ocellate," bearing a pair of prominently pigmented eyespots (Fig. 2–5) that serve primarily to sense the intensity and direction of light rather than to see a potential host. Rapid changes in light intensity, which may indicate a shadow cast by a potential host, initiate forward swimming in cercariae of *Trichobilharzia* (Feiler and Haas 1988a). Cercariae of all species treated with silver nitrate reveal a range of broadly distributed argentophilic surface organelles presumed to be sensory, but little is known of their actual functioning or their role in the life of the organism (Nuttman 1971).

The physiology of cercariae has not been studied extensively. Ercoli et al. (1985) found that addition of histamine, serotonin, or acetylcholine to the medium had no apparent effect on cercariae of *S. mansoni.* Nevertheless, histaminergic mechanisms exist in these cercariae, because specific inhibition of H1 (but not H2) histamine receptors stops locomotion and flame cell function. The cercariae, which do not feed, contain substantial reserves of glycogen (Wilson 1987).

Host Identification. As is the case with miracidia, many cercariae are not accurate in selecting only individuals of species in which they can grow and mature, but may err by entering or attempting to enter unsuitable vertebrate hosts. Cercarial dermatitis or "swimmers' itch," resulting from the penetration of the skin by inappropriate cercariae and their subsequent destruction by the host, is well known in human medicine. The cercariae of avian schistosomes such as *Trichobilharzia* or of nonhuman mammalian species (e.g., *S. spindale*) shed by freshwater pulmonate snails are most commonly involved, but cercariae from marine snails, typically *Austrobilharzia,* may also produce dermatitis ("sea bathers' eruption") in humans. Presumably the analogous situation occurs in other species of mammals and birds when they are exposed to incompatible cercariae that normally infect humans or other hosts.

Infectivity. The "infectivity" of cercariae, a term of imprecise definition, usually refers to the proportionate recovery of worms from experimentally

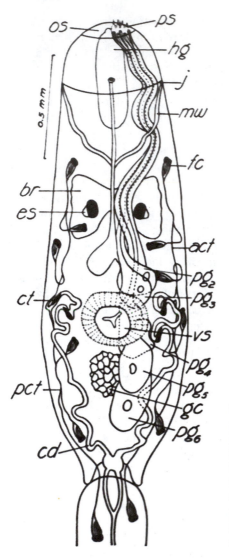

FIGURE 2–5 Ocellate cercaria of *Schistosomatium douthitti*. act, anterior collecting tubule; br, brain; cd, collecting duct; ct, ciliated part of collecting duct; es, eyespot; fc, flame cell; gc, germ cell; hg, head gland; j, line of attachment of head organ to body wall; mw, muscular wall of head organ; pct, posterior collecting tubule; pg, penetration glands; ps, penetration spine; vs, ventral sucker. *Source:* Price 1931.

exposed animals. An early study of the decrement of infectivity over time was carried out by Olivier in 1966. Groups of mice were exposed to cercariae of Puerto Rican *S. mansoni* under carefully controlled conditions. Worm recovery from these animals 42 to 44 days later was correlated with the precise age of the cercariae at the time of exposure. Infectivity remained constant for the first 3 hours, declined by 50% during the next 5 hours, and dropped to 10% at a cercarial age of 14–15 hours. Olivier (1966) considered that under field conditions the cercarial half-life is probably "only a few hours." In the same area, Stirewalt and Fregeau (1965) studied the effect of water type, temperature, cercarial density, and length of exposure on the recovery of adult *S. mansoni* from mice exposed by tail immersion. Distilled water was found unsuitable as an exposure medium, and triply distilled water even more so. Optimal conditions were defined as: 50 cercariae per exposure tube in dechlorinated tap water for at least one hour. Lwambo et al. (1987b) varied the temperature, salinity, or pH in tubes of water used to expose mice to 80 cercariae each of a Saudi Arabian isolate of *S. mansoni.* Adult worms were recovered from each mouse by perfusion after 8 weeks. Although infections occurred over a wide range in each case, worm burdens were highest in mice exposed to cercariae acclimated at 28C, at a salinity of 100 mg NaCl per ml, and at a pH of 5.4. It is not clear how well data from such artificial conditions using laboratory hosts can be extrapolated to the actual relations between free-ranging cercariae and actual hosts in the field.

In natural lotic (running-water) habitats, the flow rate controls the distance downstream at which cercariae of *S. mansoni* can infect mice. Upatham (1974) has shown that at relatively low flow rates of 10 to 40 centimeters per second, cercariae remained infective up to 97 meters downstream, but at the rapid flow rate of 75 cm.sec cercariae were infective to only 12 meters. He concluded that "cercariae swept down fast-flowing streams are unlikely to infect people having even prolonged contact with slow-flowing pools, unless these pools are very close to the infected snail colony." Presumably analogous conditions apply to infection of animals with cercariae of zoophilic schistosomes.

Using a different approach, Evans and Stirewalt (1957) determined that the infectivity of cercariae of *S. mansoni* is closely related to the physiological state of the intermediate host snail from which they are released. Individual snails were found to undergo cyclic changes during which the cercariae that they shed produced greater, or smaller, numbers of adult worms in mice exposed to them. This report may merit reexamination and confirmation.

In a functional interpretation of infectivity, Haas (1988) identified four stages of cercarial host-related interaction: attachment, remaining, directed creeping, and penetration. In the case of *S. japonicum* and *S. spindale,* Haas et al. (1987, 1990) believed that host finding depends primarily on chance contact, with cercariae responding to host stimuli only in the creeping and penetration phases of the process. Haas et al. (1990) described the temperature-dependent creeping of cercariae of *S. spindale* to warmer areas of the host body.

Attachment and Penetration

The attachment of *S. mansoni* cercariae to host skin is stimulated by contact, temperature gradient, and water currents, and mediated specifically by the presence of the amino acid arginine (Granzen and Haas 1986). However, neither this amino acid nor any other chemical enhances host attachment by cercariae of *S. spindale,* which is stimulated only by a rise in surface temperature (Haas et al. 1990). In *T. ocellata* cercarial attachment to duck foot skin disappears when the skin surface lipids are extracted, but is restored by reapplication of the lipids to the skin (Feiler and Haas 1988b). The attachment response in *S. japonicum* is independent of any chemical, thermal, or mechanical qualities of the substrate; subsequently cercariae remain on any surface solely on the basis of its hydrophobicity and migrate up a thermal gradient (Haas et al. 1987).

Cercariae are stimulated to attempt penetration in the presence of certain chemical substances. This can be demonstrated easily in a laboratory dish by rubbing a bit of oily skin secretion from one's forehead on the bottom of a glass or plastic dish and then adding water with cercariae. Within seconds some cercariae will attempt to enter into the oily area and may cast off their tails and then die unless transferred immediately to isotonic medium. Specific chemical stimuli for penetration of *S. mansoni* cercariae were studied by Haas and Schmit (1982), who incorporated 230 different chemicals into an experimental agar-based substrate. They reported that only aliphatic hydrocarbon chains with both polar and nonpolar ends were effective. At pH 7.0, saturated molecules were effective at 10 to 15 carbon atoms, but longer unsaturated molecules were also active. Cicchini et al. (1977) extracted human, mouse, and rat skin lipid fractions, finding that triglycerides, waxes, and squalenes favor penetration of *S. mansoni* cercariae into agar blocks, whereas sterols and their esters were inhibitory.

Penetration of *S. japonicum* cercariae into agarose blocks was stimulated by free fatty acids, of which dodecanoic acid was the most effective. Dog skin, which lacks free fatty acids, has no attraction to cercariae of *S. japonicum* (Haas et al. 1987). In *S. spindale,* attempts at penetration are mediated by fatty acids from cattle skin surface lipids (Haas et al. 1990).

Penetration responses of marine cercariae have been little studied. Zibulewsky et al. (1982) extracted lipids from chicken skin and incorporated chromatographed fractions or neutral lipid standards in agar in a dish containing sea water. Cercariae of *Austrobilharzia variglandis* responded most strongly to lipid from whole chicken skin, followed by free fatty acid and free sterol fractions. The phospholipid fraction was lethal to the cercariae. A whole neutral lipid standard containing cholesterol, oleic acid, triolein, methyl oleate, and cholesteryl oleate also produced a robust response.

At the moment of cercarial penetration the tail is cast off to die, and the body enters the skin of the definitive host, facilitated by secretion products of the cercarial preacetabular glands. One or more proteinases produced by these glands can hydrolyze azocoll, elastin, gelatin, laminin, fibronectin, keratin, and types IV and VIII basement membrane collagen (McKerrow et al. 1985, McKerrow and Doenhoff 1988). These substrates are all present in the stratum corneum, basement membrane, dermal extracellular matrix, and connective tissue that must be traversed if the former cercarial body is to continue its development. Working with mechanically transformed cercariae of *S. mansoni,* in vitro, Marikovsky et al. (1988) have identified two proteases of 28 and 60 kDa molecular weight that they believe to function both in penetration and in shedding of the cercarial glycocalyx. Of these, the 28 kDa component appears identical to the acetabular gland proteinase by amino terminal sequence analysis.

The Cercaria/Schistosomulum Transition*

Conversion from the Cercaria to the Schistosomulum

The conversion from cercaria to schistosomulum, a crucial and presumably vulnerable point in the life cycle, has been much studied: by 1974 there was sufficient information for Stirewalt to publish a 70-page review. The conversion cascade begins at the moment of penetration, and consists of a number of integrated structural and physiological changes. The following elements are involved in the conversion, within a span of about 3 hours:

1. loss of the cercarial tail and development of a lethal water sensitivity;
2. within the first 5 minutes microvilli about 100 to 150 nm in diameter, representing the production of some new membrane, form on the surface and then disappear within the first hour of parasitic existence; their presence seems to aid in the loss of most of the glycocalyx;
3. multilaminate vesicles migrate from underlying cell bodies through cytoplasmic connections into the surface syncytium, then pass to the outer surface and discharge, facilitating development from a single to a double bilayer outer unit membrane. Other structural changes occur; for example, Foley et al. (1988) have demonstrated that the surface membrane lipid phase changes from a gel to a liquid crystalline array. Biochemical changes accompany the morphological maturation of the schistosomular surface, and at least a dozen new glycoproteins are elaborated. Certain lectins such as concanavalin A bind to the schistosomular surface only after transformation has occurred;
4. discharge of the preacetabular glands;
5. a switch from aerobic respiration to anaerobic glycolysis;
6. acquisition of host antigens and other manifestations of immunological survival mechanisms.

The mechanism of tail loss was studied in detail by Howells et al. (1974), who cited simple mechanical trauma caused by movement of the tail "against the resistance of the secretion-fixed body." Scanning electron micrographs of freshly ruptured and healed tail-sockets are shown on Figure 2–6.

Samuelson and Stein (1989) have reported that cercariae whose tails are

---

*Note on terminology: The terms "schistosomulum" (singular) and "schistosomula" (plural) are used here in preference to the informal jargon "somule," "schistosomule," "somules," and "schistosomulae."

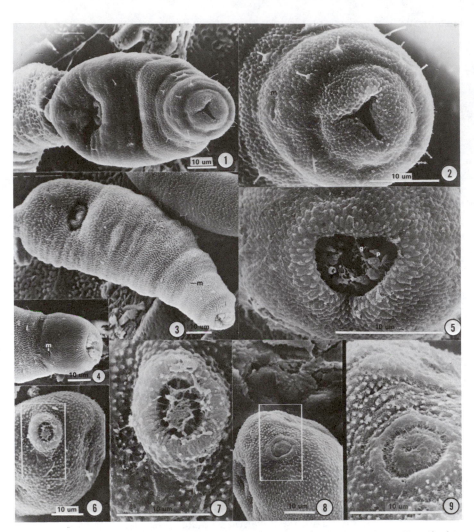

Figure 2–6 Cercaria and newly transformed schistosomula. 1, Cercaria; 2, Detail of 1, apical region and ciliated sensory papillae; 3, Schistosomulum 18 hours old. Note apical region with discharged gland cells; 4, Anterior of schistosomulum 10 minutes old with similar apical appearance; 5, Schistosomulum 10 minutes old, apical region; 6, Same, posterior end of body showing area of tail separation; 7, Detail of 6 (box); 8, Posterior view of schistosomulum 18 hours old, region comparable to that in Figure 6; 9, Detail of 8, showing healing of scar. g = gland opening, m = mouth; p = sensory papilla.

removed by shearing will demonstrate elements 2 and 3 above and transform to schistosmula after an increase in salinity. This increase may be from 18 milliosmoles, typical of pond water, to 120 mOsm of snail hemolymph; from 120 mOsm to 300 mOsm of mammalian blood; or from 18 to 400 mOsm. Therefore increasing salinity, rather than the absolute saline concentration, was considered to be the signal for transformation. Osmotic pressure per se was not the specific signal, because a mannitol solution of 300 mOsm failed to trigger the change (Samuelson and Stein 1989).

Conversion in Vitro. The need to obtain large numbers of schistosomula has led to various techniques for collection, sterilization, and decaudation of cercariae, and for cultivation of the resultant schistosomula (Brink et al. 1977, Basch 1981a, Salafsky et al. 1988). Mechanical methods of transformation include centrifuging and vortexing (Ramalho-Pinto et al. 1974), or shearing through single (Colley and Wikel 1974) or double-ended (Basch 1981a) syringe needles. Chemical methods include incubation in medium containing serum, and the use of free linoleic acid (Salafsky et al. 1988). The schistosomula derived from these varied methods, although morphologically and even ultrastructurally similar, may not be biochemically equivalent. For example, Salafsky et al. (1988) found that mechanically transformed organisms had significantly higher rates of RNA and protein synthesis than chemically transformed ones, which had higher rates of water loss and eicosanoid production. On the other hand, Thompson et al. (1984), looking at electrophysiological responses to a wide range of substances including ethanol, drugs, bioactive amines, and lectins, found remarkably little difference between schistosomula of *S. mansoni* transformed mechanically or by skin passage.

Six different transformation techniques (incubation in isotonic medium; penetration of isolated skin; stimulation with crude egg lecithin; mechanical disruption by passage through a 21-gauge syringe needle; whorling in a vortex mixer; and syringe transformation followed by incubation) and six different routes of infection (percutaneous; intradermal; subcutaneous; intraperitoneal; intravenous; and intramuscular) were assessed by James and Taylor (1976) for subsequent development of *S. mansoni* adults in mice. The largest, approximately equal, percentages of adult worms were recovered by perfusion from control mice exposed percutaneously to intact cercariae, and from schistosomula transformed by the syringe technique and injected intramuscularly.

During the first 6–8 hours after mechanical transformation, the new schistosomula lose protein, from about 2.4 to about 1.6 µg per organism, remaining constant thereafter on a short-term basis. However, the young schistosomula are metabolically very active. Protein synthesis as measured by incorporation of ^{14}C-leucine increased steadily during the first 12 hours or so and very rapidly thereafter; proteins of the tegument incorporated 2 to 3 times more isotope than proteins of the larval body (Nagai et al. 1977).

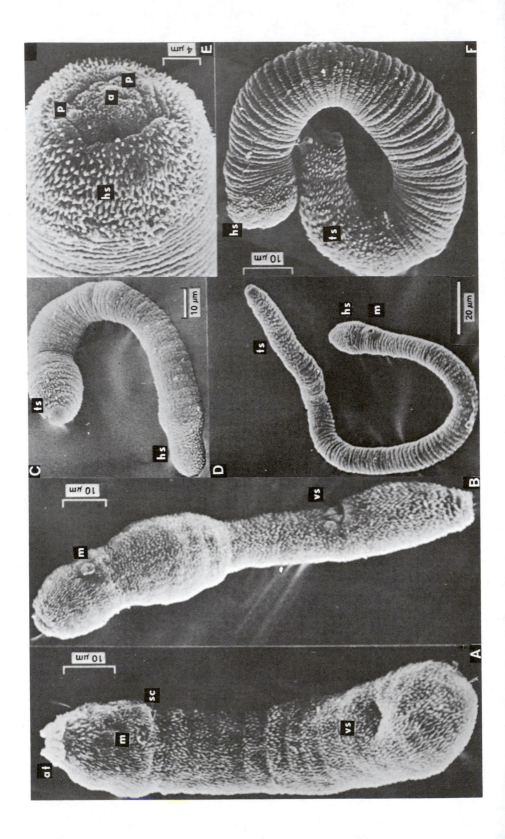

The Schistosomulum

Migration

This subject has been reviewed in detail by Wilson (1987). The route and timing of migration are known with reasonable accuracy for only a few species of mammalian schistosomes. Yolles et al. (1949) found that schistosomula of *S. mansoni* were recovered from the skin of hamsters on the fourth day and from the ears of rabbits as late as 9 days after percutaneous exposure to cercariae. Hamsters infected percutaneously first showed schistosomula in the lungs on the fourth day and in the liver on the eighth day after exposure. According to Wilson et al. (1978) young schistosomula remain in mouse skin for a mean period of 88 hours (Standard Deviation ±23 h), exiting principally by the blood vascular system for the passage to the lungs. The organisms are carried by the blood through the right side of the heart and pulmonary artery to the lung capillaries, as shown by Miller and Wilson (1980).

> After crawling through the pulmonary capillaries the worms enter the systemic circulation and are distributed randomly all over the host's body. Worms carried, by chance, to the liver, remain there. Worms carried to other organs pass through the capillary beds and into the venous circulation to return to the lungs. Worms may make several circuits of the body before eventually reaching the liver. (Lawson and Draskau 1977, about *S. mansoni*)

A series of excellent scanning electron micrographs of developing schistosomula of *S. mansoni* recovered from experimental hosts was presented by Crabtree and Wilson (1980) (Fig. 2–7), whose subsequent study (Crabtree and Wilson 1986) followed the ultrastructure of the schistosomula during their pulmonary migration. The organisms were found only in the lung vasculature on days 4 to 7 postexposure. In the lung capillaries, the fibrous layer beneath the tegument, together with some anterior musculature, breaks down in preparation for the passage through these narrow vessels. The parasites elongate and lose much of their fine body spination except for broad bands at either end, which seem to help in crawling through the cap-

FIGURE 2–7 Scanning electron micrographs of young schstosomula of *S. mansoni*. 1, Day 1 skin schistosomulum viewed from the ventral aspect. The apical tegumentary folds are still present and the entire body surface is still covered in spines; B, Day 2 skin schistosomulum viewed from the ventrolateral aspect. Elongation of the body has commenced and there is a slight reduction in the size of the spines in the mid-body region; C, day 4, from lung, viewed from the dorsal surface. Only head and tail spines are present; D, Day 7, from lung, The mouth is within a raised protruberance. E, Day, from the portal system. Anterior dorsal surface showing the retracted spineless apical area with head spines and two laterally positioned papillae; F, Day 8, from the portal system viewed from the dorsal surface. The head and tail spines are distinct and the spineless tegument has annular corrugations; a, apical area; at, apical tegumentary ridges; h, head spines; m, mouth; p, papillaa; sc, surface constriction; ts, tail spines; vs, ventral sucker. *Source:* Adapted from Crabtree and Wilson 1980.

illaries (Fig. 2–8). The young schistosomula, as small as 8 μm in diameter, fit snugly within the pulmonary capillaries and must undergo a period of adaptation before they can transit the vessels successfully (Fig. 2–9).

After passing through the lungs, schistosomula of most species that have been studied come to rest in the hepatic and portal venous complexes, except for *Schistosoma haematobium*, which goes to the vesical venules around the bladder, and the peculiar Indian cattle parasite *S. nasalis*, which goes to the mucus membranes of the nose. Almost nothing is known about the migration of the avian species. Most authors (e.g., Wheater and Wilson 1979, Wilson et al. 1978) agree that movement of *S. mansoni* from lungs to liver is entirely through the blood vascular system, as described by early workers such as Faust et al. (1934); but alternative views are considered by Miller and Wilson (1980). Schistosomula of *S. mansoni* were first seen in the portal system 6 days after infection in mice, 7 days in hamsters, and 9 days

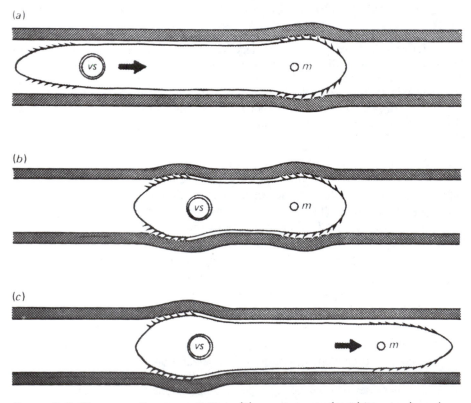

(a)

(b)

(c)

FIGURE 2–8 Diagrammatic representation of the movement of a schistosomulum along the lumen of a blood vessel, based on in vitro observations of schistosomula introduced into glass capillaries and stereoscan micrographs. (a) The elongated schistosomulum is anchored by its head spines and the posterior region is pulled up toward the head. (b) Schistosomulum at minimum length. (c) The schistosomulum is anchored by its tail spines while the anterior and midbody is extended forward. m, mouth, vs, ventral sucker. *Source:* Crabtree and Wilson 1980.

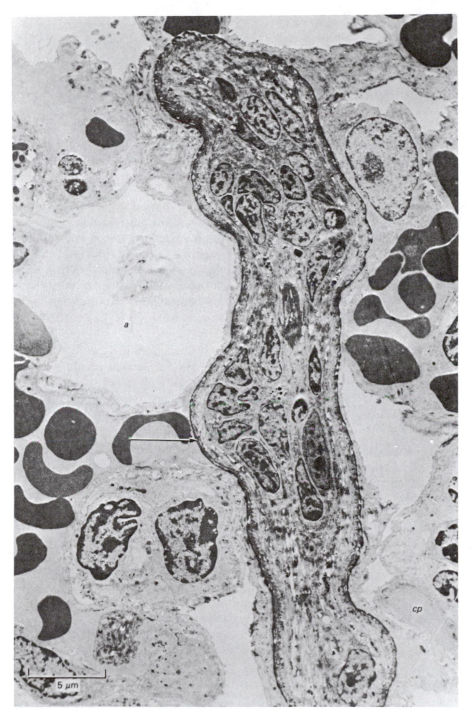

Figure 2–9 Schistosomulum of *S. mansoni* 7 days old, in capillary, longitudinal section. Transmission electron photomicrograph. Note how the body bulges into adjacent vessel on the left (arrow). The tegument is unridged and flattened against the endothelium. a, alveolus; cp, capillary. *Source:* Crabtree and Wilson 1986.

in rats. In mice and rats migration continued to about day 20, but in hamsters new schistosomula of *S. mansoni* arrived in the portal system up to 40 days after exposure (Miller and Wilson 1980). It is likely that within a few days after their arrival in the liver, schistosomula lose their ability to migrate extensively. In CBA/Ca mice Kamiya et al. (1987) reported that 84% of entering parasites migrated to the lungs, but up to half of these did not move on to the liver. Overall, 58% of parasites were lost in a series of immunologically naive mice. In mice previously exposed to irradiated cercariae, 83.5% of parasites were lost before going to the liver, and only 22% of entering parasites were recovered as adults.

Despite the many studies of schistosomular migration there is no absolute "gold standard" for staging of development, in part because of genetically determined host anatomical variation. Mitchell (1989) has demonstrated that 20-micron latex beads injected into the portal system of Balb/C mice are trapped efficiently in the liver, whereas in 129/J and C57BL/6 mice most of the beads proceed to the lungs. Inherent peculiarities of portal system configuration are of great significance in determining relative susceptibility or resistance to schistosomes and the subsequent course of parasite development.

An elaborate experimental system using tissue autoradiography to follow the movements of radioactively labeled 7-day-old schistosomula was described by Wilson and Coulson (1986). These authors found that the distribution of injected schistosomula to systemic organs, following lung transit, "was proportional to the factional distribution of cardiac output." The population of schistosomes eventually to reside within the portal system was assembled during 2 to 3 circuits around the pulmonary-systemic blood vessels. Georgi et al. (1986), in contemporaneous studies by similar methods, observed attrition of schistosomula of *S. mansoni* and *S. haematobium* from all organs other than the liver in the mouse. In *S. mansoni* schistosomula peaked in the lung on day 6 postexposure and began accumulating in the liver on day 7, stabilizing at the final number about day 14. Georgi et al. (1986) disputed the various reports of Wilson and colleagues that schistosomula circulate more or less randomly and repeatedly through the pulmonary-systemic blood vessels, considering it more likely that the lung-to-liver migration is specifically directed through connecting vessels. This point requires clarification.

For *S. japonicum* Cort (1921a) provided a useful set of drawings of developmental stages (Fig. 2–10). Miyagawa and Takemoto (1921) in Japan described the route from the skin to the portal vein in mice. They concluded that after leaving the skin the young worms generally invaded blood capillaries, later passing through the right side of the heart and lungs, then via the circulatory system to the hepatic veins. For an isolate of *S. japonicum* from China, He and Hang (1980) followed the post-cercarial development, finding that schistosomula remained in the skin not more than 24 hours, reached the lungs on day 2, and the liver on day 3 after exposure. They remained in the liver for 8 to 10 days, passing to the portal and mesenteric veins as early as day 11. Moloney and Webbe (1983) studied *S. japonicum* (originally from Zhejiang Province, China) in CBA white mice. They found peak passage through the lungs at days 5–6, and no significant increase in worms recovered from the portal system after 2 weeks of infection.

Schistosomula of *S. bovis* begin to arrive in the lung 4 days after infection, and in the liver 3 days later (Lengy 1962b). In sheep, Kruger et al. (1969) made a special study of the route of migration, concluding that schistosomula moved against the bloodstream via the pulmonary artery, right ventricle and atrium, posterior vena cava, and hepatic vein into the liver. The absence of schistosomula from the left heart, pulmonary vein, and aorta caused these authors to reach a conclusion contradictory to that of many previous authors working with laboratory animals infected with the schistosomes of man.

Cort (1921a) established 3 periods of growth: (1) from penetration of the cercaria until the young worm reaches the blood vessels of the liver, during which there is little increase in size; (2) from the time of reaching the liver until sexual maturity, when growth is rapid and concentrated; and (3) from sexual maturity to maximum size, when growth is slow and extended in duration. Although referring specifically to *S. japonicum,* this general outline appears suitable for all species of schistosomes.

Growth and development of portal system worms have been followed from schistosomula to adults by many investigators. The stated time of appearance of certain maturation landmarks is indicated on Table 2–1, but growth curves and verbal descriptions are more difficult to summarize and the studies cited should be consulted for details. A set of general staging standards for development of *S. mansoni* in the mouse was defined by Clegg (1959):

Stage O. Cercaria.

Stage 1. Lung larva. 7 days.
> Although the gut and excretory system develop slightly and the body is elongated, larvae in the lungs remain the same total size as cercariae. Since mitosis cannot be demonstrated it appears that growth has not started.

Stage 2. Gut development. 15 days.
> Larvae enter the portal vessels of the liver on the 8th day where they begin to feed on blood and grow. The gut caeca grow around the ventral sucker and join together some distance posterior to it on the 15th day. Mitoses can be counted easily. Males are larger than females and show more mitotic activity.

Stage 3. Organogeny. 21 days.
> Sex organs begin to develop at the end of the third week. In males two testes can be distinguished without internal differentiation; in females a narrow uterus is visible.

Stage 4. Gametogeny. 28 days.
> Eight testes have developed with sperm in the anterior two or three. Female ducts have formed and a small ovary is present. Copulation begins.

Stage 5. Appearance of egg-shell protein. 30 days.
> Vitelline globules can be stained by diazonium salts.

(*continued on p. 54*)

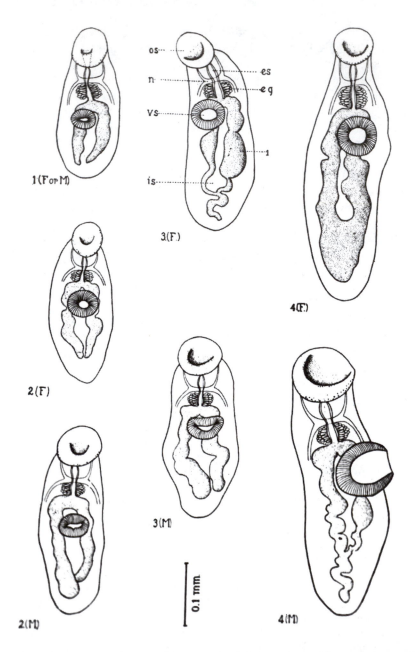

FIGURE 2–10 Developing schistosomula of *S. japonicum*, from experimentally infected mice. All figures drawn with the aid of a camera lucida. 1. From liver, 10 days Sex uncertain; 2, 3, 4 (F), from liver, 12 days, females; 2, 3 (M), from liver, 12 days, males; 4 (M), from liver, 16 days, male; 5, 6, 7, 8 (F), from liver, 14, 16, 18, and 19 days,

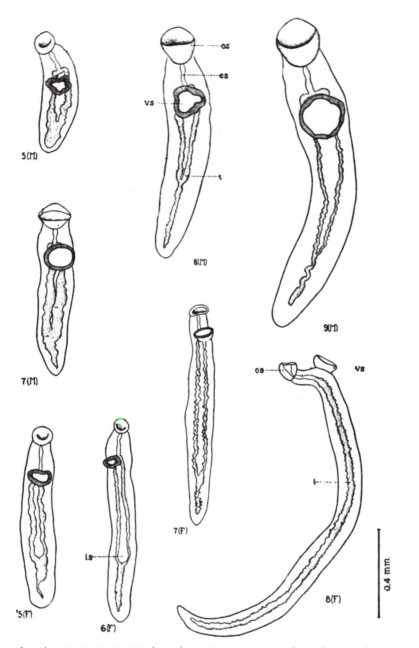

females; 5, 7, 8, 9 (M), from liver, 14, 14, 16, and 16 days, males. eg, esophageal gland; es, esophagus; i, intestine; is, intestinal space in female where ovary develops; n, central nerve mass; os, oral sucker; vs, ventral sucker. Consult original for complete dimensions and detailed description. *Source:* Cort 1921.

TABLE 2–1 Schistosome Maturation Landmarks

Species/Origin	Host	Time After Infection	Observation	Reference Note
S. mansoni				
Puerto Rico	Mouse	25 days	Copulation	1
		28 d	Eggshell protein	
		30 d	Oviposition	
Puerto Rico	Mouse	30 d	Eggs in uterus of mated female	2
		8–10 wks	Complete maturity	
Puerto Rico	Mouse	5 w	Copula	3
		5 2/7 w	Spermatids in M seminal vesicle	
		5 4/7 w	Vitelline glands functional; sperm in seminal receptable; egg in ootype	
		6 w	Full maturity, Male and Female	
Puerto Rico	*In vitro*	5 d	Black pigment in gut	4
		11–12 d	Fusion of gut ceca	
		7th w	Pairing	
Brazil	Mouse	21 d	Sexual development begins	5
		28 g	Pairing; Gamete development begins	
		35 d	Oviposition	
Brazil	Rodents	5 wk	Egg in uterus; Sexual organs well developed	6
S. mansoni				
Liberia?	Mouse	30 d	First pairs	7
			Mature and functional paired F; sperm in oviduct	
—	Mouse	36 d	Eggs in tissues	8
—	Mouse	7 d	Lung larva (optimum)	9
		15 d	Gut development	
		21 d	Organogeny	
		28 d	Gametogeny	
		30 d	Eggshell protein	
		35 d	Oviposition	
S. haematobium				
Ghana	Hamster	29 d	Copulation	10
		45 d	Eggshell protein	
		53 d	Fluorescent material in vitelline Duct	
		60 d	Egg production	
Sudan	Hamster	29–31 d	First signs of ova and sperm	11
		31 d	Pairing	
		57 d	Eggshell protein	
		63 d	Poor eggs	
			Good eggs	

TABLE 2–1 (continued)

Species/ Origin	Host	Time After Infection	Observation	Reference Note
Egypt	Mouse	12 w	84% mature	12
		14 w	100% mature	
	Hamster	9 w	4% mature	
		10 w	100% mature	
Iran	Hamster	55 d	First eggs (also Ghana, Mauritius)	13
Israel?	—	35–42 d	Sexual maturity	14
S. japonicum				
China	Ch hamster	26 d	Spermatids in unisexual males	15
		40 d	Paired males fully mature	
		45 d	Maturing sperm in unisexual males	
		54 d	Unisexual males fully mature	
		65 d	Paired females fully mature	
China	Mouse	15–19 d	Mating	16
		19 d	Sperm in testes	
		20 d	Ova in ovaries	
		21 d	Vitelline cells	
		22 d	Eggshell protein; egg in uterus	
		29 d	Oviposition	
	Rabbit	20 d	Pairing (possibly earlier)	
		21 d	Egg in uterus	
Japan	Mouse	19 d	First copulation	17
Japan	Mouse	6 d	Sex differentiation	18
		17 d	Mating	
	Dog	28 or 30	Maturity	
Japan	Dog	23	Maturation	19
	Calf	26	Maturation	
Japan?	Mouse	20 d	Pairing (possibly earlier)	20
		25 d	First egg deposition	
Philippines	Mouse	20 d	Few eggs	21
S. bovis				
Corsica	Mouse	63 d	Eggs in tissues	22
Israel	Mouse	19 d	Sexual differentiation	23
		33 d	Copulation	
		42 d	Egg in uterus	
S. margarebowiei				
Zambia?	M, H, G	20 d	Organogeny	24
		25 d	Gametogeny; pairing	
		30 d	Eggshell formation	
		33 d	Oviposition (hamsters and mice)	
		35 d	Oviposition (gerbils)	
S. mekongi				
Cambodia	Mouse	26 d	Mating (copula)	25
		30 d	Egg in uterus	

TABLE 2–1 Schistosome Maturation Landmarks (*continued*)

Species/ Origin	Host	Time After Infection	Observation	Reference Note
S. douthitti				
Michigan	*Peromyscus*	11 d	Shell material in vit gland of paired females	26
		12 d	Pairing; sperm in males	
Michigan	*Peromyscus*	10 d	Sexual maturity of females	27
		12 d	Sexual maturity of males	
		13 d	Mating	
		14 d	Sperm in seminal receptacle of females	
Heterobilharzia americana				
Louisiana	Mouse	17 d	Female genital primordium elongated	28
		21 d	Uterus well defined; testes first noted	
Trichobilharzia ocellata				
Canada	Duck	4 d	Testes present in males	29
		6 d	Sperm and gynecophoral canal present	
		7 d	Maturity in males	
		9 d	Egg laying by females	

Reference notes: 1, Burden and Ubelaker 1981; 2, Armstrong 1965; 3, Moore et al. 1955; 4, Basch 1981a; 5, Michaels 1970; 6, Lutz 1919; 7, Vogel 1941b; 8, Brumpt 1930; 9, Clegg 1959; 10, Burden and Ubelaker 1981; 11, Smith et al. 1976; 12, Moore and Meleney 1954; 13, Wright and Knowles 1972; 14, LeRoux 1949; 15, Severinghaus 1928; 16, He and Yang 1980; 17, Cort 1921a; 18, Faust and Meleney 1924; 19, Fujinami and Nakamura, 1909; 20, Vogel 1942a; 21, Voge et al. 1978a; 22, Brumpt 1930a; 23, Lengy 1962b; 24, Ogbe 1983; 25, Voge et al. 1978a; 26, Short 1952a; 27, El Gindy 1951; 28, Lee 1962; 29, Bourns et al. 1973.

Stage 6. Oviposition. 35 days.
Vitelline glands are well developed and the first eggs are produced. Sperm is plentiful in all testes. Paired schistosomes have usually migrated from the portal vein into the mesenteris vessels.

Recorded times until egg production are summarized on Table 2–2.

Lutz (1919) found that 3 weeks after infection of rodents, his Brazilian *S. mansoni* occurred in equal numbers in the liver and the mesenteric veins "which are generally very congested." In their classic study in Puerto Rico, Faust et al. (1934) pointed out that growth and development of the schistosomula beyond the very earliest stages can take place only in portal blood; moreover, attainment of maturity of male and female worms is practically synchronous. Their representation of the stages of growth has served as a standard of comparison for later works (e.g., Basch 1981a). An abbreviated but more modern illustration of key growth stages of *S. mansoni* is presented by Clegg (1965b). Tables of linear dimensions and growth curves of unisex-

TABLE 2–2 Timing of Egg Production

Species/ Origin	Host	Days to First Egg	Observations	Reference Note
S. mansoni				
St. Lucia	Mouse	32	In tissue	1
Puerto Rico	Mouse	33	In tissue	
Puerto Rico	*Rhesus*	36–39	Feces. Max. No. @ 50–80 days	2
Puerto Rico	Human	42	Feces. Accid. lab infection	3
Puerto Rico	Mouse	53	In feces	4
Puerto Rico	Hamster	43	In feces	5
Puerto Rico	*Rhesus*	36	In feces	6
Puerto Rico	Mouse	42	In feces	7
Puerto Rico	Hamster	40	In feces	
Brazil	Mouse	36	In tissue	8
Brazil	Guinea pig	2.5 mo	Feces: small numbers	9
Tanzania	Mouse	42	In tissue	10
Tanzania	*Rhesus*	39	In feces	11
—	Mouse	45	In feces	12
Tanzania	Mouse	49	In feces	13
Tanzania	Hamster	49	In feces	
Sierra Leone	*Cercopithecus*	42	In feces	14
Egypt	Mouse	5 wk	In feces	15
Egypt	Mouse	41	In feces	16
Egypt	Hamster	38	In feces	
Egypt	*Rhesus*	57		17
S. rodhaini				
Zaire	Mouse	30		18
Ruanda-Urundi	Mouse	30		
Uganda	Mouse	31		
Uganda	Hamster	33		
S. haematobium				
Sudan	Hamster	ca. 70		19
Sudan	Hamster	70	In feces	20
S. Africa	Hamster	ca. 70		
Egypt	Hamster	65		21
Egypt	various	7–18 wk	Depends on host	22
Egypt	*Rhesus*	91	In feces	23
Tanzania	Hamster	70		
S. Africa	Hamster	75		24
Sudan	Hamster		Malformed eggs at 63 days	25
Sierra Leone	*Cercopithecus*	101	In feces	26
S. intercalatum				
	Hamster	45	50% have eggs in uterus	27
		48	80% have eggs in uterus	
		60	>90% have eggs in uterus	

55

TABLE 2–2 Timing of Egg Production (*continued*)

Species/ Origin	Host	Days to First Egg	Observations	Reference Note
S. incognitum				
India	Pig	36	In feces	28
Java	*Rattus exulans*	>50	Also mouse	29
S. leiperi				
Zambia	Hamster	40	No eggs in uterus	30
		49	In feces	
		50	86% have eggs in uterus	
S. mattheei				
Transvaal	Hamster	50		31
Zimbabwe	Calf	6–7 wk	In feces	32
S. margarebowiei				
Botswana	Hamster	28	First eggs seen	33
		33	In feces	
		35	95% of females have eggs	
	Sheep	38	In feces	
S. nasalis				
India	Goat	5 mo	Nasal discharge	35
	Calf, female	7 mo		
	Calf, male	11 wk		
	Buffalo calf	12 wk		
India	Calf	84	Nasal discharge	
S. douthitti				
Michigan	*Peromyscus*	12–13	1 day later in unisexual females	36
Michigan	*Peromyscus*	28	Immature eggs	37
		30	Mature miracidia	
Heterobilharzia americana				
Texas	Dog	72		38
S. japonicum				
Philippines	Guinea pig	4–5 wk	Eggs in liver at 5 wk	39
	Mouse	23–27		
	Mouse	35	In feces	40
	Dog	35	In feces	
	Hamster	38	In feces	41
	Mouse	32	1 worm pair	42
China	Various	39–41	In feces	43
	Mouse	30	First eggs in host tissue	44
	Mouse	25		45
	Mouse/Rabbit	25–26		46
	Dog	29	Had 10,000 worms. Died at 33 days	47
	Mouse	36	In feces	48
	Mouse	41	1 worm pair	49
	Guinea pig	34		50

TABLE 2–2 (continued)

Species/ Origin	Host	Days to First Egg	Observations	Reference Note
Taiwan	Various	40–77	In feces	51
	Various	36–40	In feces	52
	Mouse	38	1 worm pair	53
	Hamster	39	In feces	54
Japan	Various	35–39	In feces	55
	Various	31–37	In feces	56
	Mouse	35	1 worm pair	57
S. mekongi				
Laos	Mouse	35	longer than *S. japonicum*	58
Trichobilharzia ocellata				
Canada	Duck	9		59
Trichobilharzia brevis				
Malaysia	Duck	11	First hatched miracidia	60

Reference notes: 1, Anderson and Cheever 1972; 2, Cheever and Powers 1969; 3, Faust et al. 1934; 4, Kloetzel 1967b; 5, Moore and Sandground 1956; 6, Nelson and Saoud 1968; 7, Saoud 1965; 8, Anderson and Cheever 1972; 9, Lutz 1919; 10, Anderson and Cheever 1972; 11, Nelson and Saoud 1968; 12, Berg 1953; 13, Saoud 1965; 14, Gordon et al. 1934; 15, Kuntz 1961a; 16, Saoud 1965; 17, Standen 1949; 18, Fripp 1968; 19, James and Webbe 1975; 20, Smith et al. 1976; 21, James and Webbe 1973; 22, Kuntz 1961b; 32, Standen 1949; 24, Wright and Ross 1980; 25, Smith et al. 1976; 26, Gordon et al. 1934; 27, Wright et al. 1972; 28, Sinha and Srivastava 60; 29, Carney et al. 1977; 30, Southgate et al. 1981; 31, Wright and Ross 1980; 32, Lawrence JA 1973a; 33, Southgate and Knowles 1977; 34, Dutt and Srivastava 1968; 35, Rao 1938; 36, Short 1952a; 37, Price 1931; 38, Goff and Ronald 1981; 39, Bang and Hairston 1946; 40, Chiu and Kao 1973; 41, Hsü and Hsü 1958b; 42, Hsü et al. 1960; 43, Chiu and Kao 1973; Moloney and Webbe 1983; 45, Vogel 1942a; 46, Yen et al. 1974; 47, Faust and Meleney 1924; 48, Hsü et al. 1958; 49, Hsü et al. 1960; 50, Vogel and Minning 1947; 51, Chiu and Kao 1973; 52, Hsü and Hsü 1958b; 53, Hsü et al. 1960; 54, Moore and Sandground 1956; 55, Chiu and Kao 1973; 56, Hsü and Hsü 1958b; 57, Hsü et al. 1960; 58, Voge et al. 1978; 59, Bourns et al. 1973; 60, Basch 1966.

ual and paired male and female worms of West African origin may be found in Vogel (1941b); growth curves comparing male and female *S. mansoni* (Puerto Rico) and *S. haematobium* (Ghana) are given by Burden and Ubelaker (1981). As an indication of growth, the body weight of schistosomula of *S. mansoni* has been determined by several authors. Lawson and Wilson (1980) followed the wet weight from 0 to 24 days after infection, and also presented data on nitrogen content and oxygen consumption during this period. The dry weight and glycogen content of schistosomula between 30 and 100 days of age was studied by Lennox and Schiller (1972). They found that the weight of males increased 94-fold between 20 and 90 days, while the weight of females increased about 34 times. At 50 days males had a mean weight of 91 µg, and a maximum, at 90 days, of 176 µg. Females attained a mean weight 26% of that of males.

During this period of growth and development profound changes occur in all organ systems with maturation of the tegument (Fig. 2–11) and the intestinal epithelium (Fig. 2–12).

For members of the *S. haematobium* group, growth is very much slower than for *S. mansoni*. Wright and Bennett (1967a) showed that growth of worms from Durban was slow until 90 to about 140 days after infection, but then became rapid during the next 40 days. The largest individuals of both sexes were recovered at 160 days of age, demonstrating the necessity of maintaining infected animals for long periods; indeed, the length of female

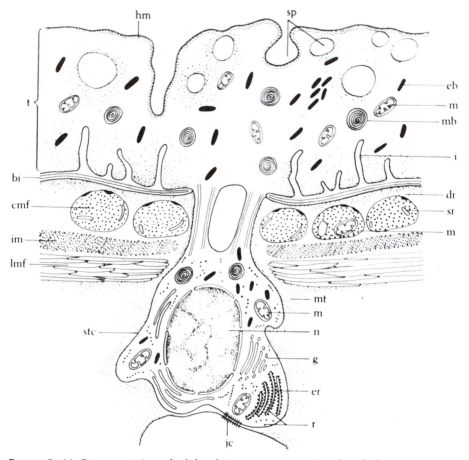

FIGURE 2–11 Representation of adult schistosome tegument and underlying structures. bi, basal lamina; cmf, circularly arranged muscle fibers; dr, peripheral dense region of muscle fiber; eb, elongate body; g, Golgi region; hm, heptalaminate outer membrane; i, invagination of basal membrane; im, interstitital material; jc, junctional complex with parenchymal cell; lmf, longitudinal muscle fiber; m, mitochondrion; mb, membranous body; mt, microtubule; n, nucleus; r, ribosomes; sp, surface pits; sr, sarcoplasmic reticulum; stc, subtegumental cell; t, tegument. *Source*: Basch and Samuelson, 1990, adapted from Hockley 1973.

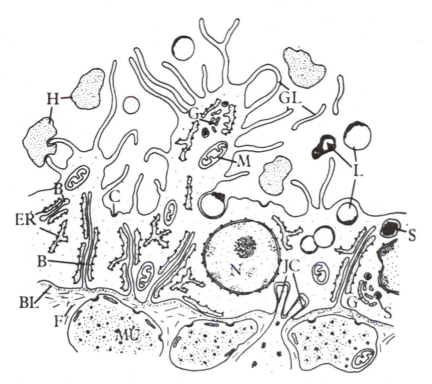

FIGURE 2–12 Diagram of the fine structure of generalized gut epithelium of *S. mansoni*. B, basal invagination; BL, basal lamina; C, micropinocytotic ceveolus; ER, endoplasmic reticulum; F, fibrous layer; G, Golgi complex; GL, gut lamella; H, hemoglobin; JC, junctional complex; L, lipidlike droplet; M, mitochondrion; MU, muscle; N, nucleus; S, dense secretion body. *Source:* Morris 1968.

worms doubled between days 120 and 160. The retarded growth of males retained in the liver was also described by Wright and Bennett (1967a). Growth of *S. haematobium* of various geographic isolates was described by Wright and Bennett (1967b), who found that worms of Iranian origin grew most slowly and those of Mauritian origin most rapidly under standardized laboratory conditions.

The growth of male and female *S. haematobium* of Iranian and Nigerian origin was compared by Webbe and James (1971a, b). Iranian worms were larger and continued to grow from 75 to 170 days after infection, while Nigerian worms remained stable after 80 or 90 days of development. In a similar way the growth of worms of Sudanese and South African origin was compared by James and Webbe (1975), who found the former to be consistently larger. Although the evidence to date is largely anecdotal, it is likely that schistosome growth characteristics are genetically determined. The relationships are complex, however, as demonstrated by Wright and Ross (1980), who found that hybrids between *S. haematobium* and *S. mattheei* grew more rapidly, matured earlier, and produced more eggs than either parental stock.

Growth of *S. japonicum* has received scant attention in recent decades. Early parasitologists such as Katsurada (1904) established general dimensions of the parasite and suggested that the size of the host influences the size of the schistosomes. Cort (1921a) recorded the range of sizes of *S. japonicum* at specific intervals after infection (Fig. 2–10). Faust and Meleney (1924) plotted growth curves for male and female *S. japonicum* up to 80 days of age in mice, and Severinghaus (1928) showed growth curves for unisexual and paired male and female worms.

Information on growth and development patterns is scarce for other species of mammalian schistosomes. Growth curves in hamsters for *S. mattheei* (Transvaal) and for *S. intercalatum* (Lower Guinea) were constructed by Wright et al. (1972). The latter species was also studied in mice by Bjorneboe and Frandsen (1979), who compared the growth patterns of male and female worms originating in Cameroon and Zaire. Tewari and Singh (1979) looked at *S. incognitum* from India, providing data on lengths of males and females at 2, 4, 8, and 12 weeks after infection in mice. For *S. margrebowiei*, growth curves of paired males and females indicate that growth in hamsters continues to at least 60 days: the longest individual female measured 33.8 mm, which may represent the largest mammalian schistosome (Southgate and Knowles 1977). Ogbe (1983) illustrated key developmental stages of the same species and presented a table comparing rates of development of five species of mammalian schistosomes. Growth curves for adult Zambian *S. leiperi* in hamsters were shown by Southgate et al. (1981) from 40 to 100 days after infection. Females grow more rapidly than males and growth apparently continues past 100 days.

Little information is available about mammalian schistosomes of other genera. For *Schistosomatium douthitti* Short (1952a) reported that growth continues until about 40 days after infection. From this point onward, males remain the same length, but paired females become progressively shorter until at 383 days of age they are about the same length as females of comparable age grown without males. Comprehensive growth data are presented in that paper.

Site and Mate Finding; Migration to the Mesenteric Veins

The requirement for mate-finding is stringent in most schistosomes, and has stimulated various investigators to search for specific chemotactic substances effective in communication between male and female worms. The subject of chemical communication in platyhelminthes has been reviewed by Bone (1982). Armstrong (1965) infected mice with mixed males and females of *S. mansoni*, *Schistosomatium douthitti*, and/or *Heterobilharzia americana*, finding almost random pairing during early phases of the mating process. He concluded that in the mouse, at least, mating occurred by thigmotaxis and trial and error: "It would seem that the schistosomes are attracted not so much to one another, initially, as they were to a rendezvous site within the host" (Armstrong 1965).

Mate-finding almost always occurs in the liver or perhaps in the proximal portal vein, and few, if any, *S. mansoni* are found outside the liver within the first 26 days after infection (Standen 1953a). Standen found pairing to occur on day 30 in mice exposed simultaneously to both sexes of cercariae, and on day 23 after giving male cercariae to mice infected earlier with females. Pairing in *S. mansoni* was found by Wilson (1987) to start between 28 and 35 days after infection.

In the strange schistosome of Australian crocodiles reported by Platt et al. (1991) males were already paired when 250 μm in size!

Evidence in favor of chemoattractants was offered by Shirazian and Schiller (1982). In their study, pairs of *S. mansoni* were taken from mice and carefully separated. Individual male and female worms, consisting of original partners or exchanged partners, were placed in small glass tubes in appropriate culture medium and observed for pairing or re-pairing. It was reported that within the first 24 hours the original partners paired faster than the exchanged worms, with a statistical probability of <0.001. Similar results were obtained in a different experimental protocol by Eveland et al. (1982) who found migratory attractiveness greater between individuals from the same than from different pairs; greater attraction was found between worms of different sexes than of the same sex. Eveland et al. (1983) found that neither males nor females of *S. mansoni* were attracted to worm pairs, to two females, or to two males, although heterosexual attraction occurred to single worms: this was interpreted as a possible threshold effect. In a related experiment Imperia et al. (1980) concluded that male *S. mansoni* emit pheromones attractive to females, but inexplicably single males were found more attractive than groups of three.

The presumed age at onset of chemoattraction was studied by Eveland and Haseeb (1989), who worked with juvenile worms perfused from mice before pairing. They reported that 20- and 21-day-old worms of opposite sex did not attract each other in vitro, whereas 23 to 28-day-old worms did do so, and subsequently paired. The demonstration of onset of chemoattraction in vitro was interpreted as corresponding to the comparable events in vivo.

Evidence that argues against the presence of chemoattractants was presented by Armstrong (1965), who found that mating was almost random in mixed infections of *S. mansoni*, *Schistosomatium douthitti*, and *Heterobilharzia americana* in mice. Similarly, Southgate et al. (1982) concluded that no specific mate recognition exists for *S. haematobium* and *S. intercalatum*. In mixed infections, males of *S. haematobium* were better at pairing with females of either spaces than were males of *S. intercalatum*.

The well-known predilection of *S. mansoni* for the posterior branches of the superior and inferior mesenteric veins, of *S. japonicum* for the anterior branches of the superior mesenteric veins, and of *S. haematobium* for the vesical plexus (Faust et al. 1934) shows the capability of these organisms to locate their respective final microhabitat within the host. Adults of *Orientobilharzia turkestanicum* are found in ruminants only in the veins of the duodenum, whereas schistosomula of *S. bovis* distribute themselves evenly in the mesenteric veins (Massoud 1973b), as shown on Figure 2–13. The remarkable concentration of *S. nasalis* in the veins of the turbinate folds of the mucous membrane of the nose of cattle (Rao 1938) is especially noteworthy. According to Farley (1971) a species of *Trichobilharzia* resides in a similar site in certain birds. These are clearly adaptations for discharge of eggs from the host. The means by which the worms can identify the appropriate division of the venous system are not understood, and generalities such as "instinctive behavior" or "chemoattraction" are commonly invoked as an indication of our ignorance.

Although unisexual male worms can and do migrate from the liver to the mesenteric veins, pairing was considered by Standen (1953a, b) to provide the main stimulus for migration of *S. mansoni* into the portal vein and beyond. The migratory impulse of the male is considerably stimulated by the presence of females: in an interesting experiment Standen (1953b) superinfected mice having unisexual male *S. mansoni* with female cercariae. It was found that the male worms tended to wait for the females to achieve a certain stage of development before migrating from the liver, although they were capable of migration throughout the whole period.

Young males were found very active at 24–26 days of age; if placed into

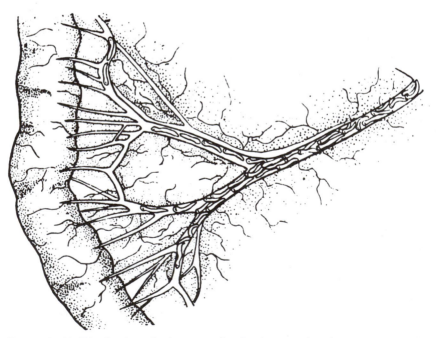

FIGURE 2–13 Distribution of schistosomula of *S. bovis* within the mesenteric veins of a rabbit. *Source:* Brumpt 1930a.

a petri dish "they will move across it, using their suckers in leech-like fashion. Young females do not possess this ability . . ." (Standen 1953a). The male was described as the active partner, "hauling a passive female." More than 30 years earlier Manson-Bahr and Fairley (1920) had made an almost identical observation: "The male assumes the chief part in this progress, the female being carried in a purely mechanical manner, in the gynaecophoric canal." Similar behavior occurs in *S. mansoni* cultured in vitro from cercariae and fed on human erythrocyte ghosts (Basch 1984); worms fed on intact erythrocytes never showed such activity.

The first evidence of migration of *S. mansoni* to the mesenteric veins in rats was found by Faust et al. (1934) at about 546 hours post infection; by 720 hours many unmated adolescent worms were found in the posterior branches of the superior mesenteric vein in the rat and rabbit. It is not clear whether this represents aberrant behavior in nonpermissive hosts, for no sexually mature females were recovered from these animals even after 40 days of infection. In *S. japonicum* Ito (1954) reported 30% of worms in the mesenteric veins of rabbits, guinea pigs, and mice in the second week post-infection, and 80% by the seventh week.

It is a common finding in unisexual or unbalanced infections that the unpaired individuals remain in the liver or hepatic portal sinus. In my own experience, primarily with *S. mansoni,* it is rare to find an unpaired male in the mesenteric veins, and unisexual females are markedly absent. However, Foster et al. (1968) reported on the basis of differential perfusion that in this species some immature, unpaired female worms normally migrate to the mesenteric veins. In heavy infections of mice with *S. bovis* (normally a parasite of ruminants), Lengy (1962b) found paired and unpaired adults as well as schistosomula throughout the portal system.

Standen (1953a) believed that an extended stay of unpaired males in the liver had positive survival value, for in nature the host would be subject to reinfection, and supernumerary males, already resident in the liver, would be available for pairing with newly acquired females. The presence of attractants to lead schistosome pairs into the mesenteric (or other specific) veins has not been much investigated. A possible clue was offered by Awwad and Bell (1978), who dialyzed human feces and added small amounts of the dialysate to culture medium in a linear track apparatus. Worm pairs introduced in the center of a tube could migrate to the dialysate end or to normal culture medium. It was found that 23 of 30 pairs moved towards the bait, supporting a hypothesis that feces contains a chemoattractant specific for pairs. Of single males or females only 12 of 30 migrated towards the bait end.

Armstrong (1965) found that homosexual male pairs of *S. mansoni* remained in or near the liver. He speculated that the mere presence of a body in the gynecophoral canal is an insufficient stimulus for migration to the site of egg deposition, but that the body in the canal must be a female, a finding corroborated by Smith and Chappell (1990). On the other hand, *Schistosomatium douthitti* of both sexes may migrate singly to the mesenteric venules, and single males and females of *Heterobilharzia americana* are occasionally found there (Armstrong 1965). In *S. douthitti* there is a tendency for unisexual females to linger in the liver: 14 days after infection 77% of unisexual female worms are found there, whereas only 2% of females from paired infections remain so long (Short 1952a).

In a chemotherapy trial with *S. mansoni* in mice, Khayyal (1964) found that a single injection of tartar emetic caused the sexes to separate and shift to the liver within 2 hours; at about 8 hours, when the worms were beginning to return to the mesenteric veins, there was a preponderance of males in the portal and mesenteric veins, leaving the respective females in the liver. After a week, the worms were again paired in the mesenteric veins as before. It was not clear whether the recovered females traveled independently to repair with males or whether the males traveled back to the liver to retrieve their mates. Almost the same phenomenon was reported a decade later by Foster and Cheetham (1973), who treated *S. mansoni*-infected mice with oxamniquine. Worms shifted from the mesenteric veins to the liver over a period of about 6 days. Males were retained in the liver and after 14 days 88% were dead. However, during the second week after treatment there was a return shift back to the mesenteries by the surviving, unpaired females. How this shift was accomplished against the blood flow is unexplained.

Distribution of Adult Worms

Faust et al. (1934) suggested that there is considerable movement by *S. mansoni* from one site to another within the mesenteric vessels. This was confirmed by Pellegrino and Coelho (1978), who exposed individual 2- to 6-day-old mice to exactly three cercariae of *S. mansoni,* yielding 20% with a single pair of worms. Animals were sacrificed from 55 to 236 days after infection, and the entire length of the intestinal tract was examined for schistosome eggs by the oogram method. Eggs were found essentially throughout, demonstrating a "remarkable wandering capacity of schistosome pairs along the extensive and complex venous system draining the intestinal tract in mice."

Later studies by a similar method (Valadares et al. 1980, 1981a, b) showed that the propensity to wander varied among worms isolated from different geographic sites within Brazil. It was also shown that individual females varied in this characteristic, even within a single geographic isolate (LE' from Belo Horizonte, Brazil). A comparison of the migration of single pairs from Japanese, Taiwanese, and Chinese isolates of *S. japonicum* has been done in the same manner by Cheever and Duvall (1982). Whereas favored sites for oviposition differed to some extent, eggs were found broadly distributed in the intestines. The presence of larger focal lesions suggested that worm pairs, particularly of Japanese origin, are attracted to existing sites of egg deposition, perhaps by substances released into the mesenteric blood from eggs or the surrounding inflammatory exudate.

The actual movement of adult *S. mansoni* within vessels was seen

RELATIONS WITH VERTEBRATE HOSTS / 65

through a high resolution television system modified for in vivo microscopy (Bloch 1980). Both single and paired worms moved constantly, usually in an undulating fashion against the blood flow. The worms became sufficiently elongated to enter smaller vessels, such as the mural venules at the mesenteric-intestinal junction.

Migration to Unusual Sites

Adult schistosomes have been reported many times from the human genitalia, particularly the vagina. Renaud et al. (1989) described a village in Niger with vaginal schistosomiasis caused by *S. haematobium*, and de Gentile et al. (1988) found the same species in the epididymis. In Nigeria Obasi (1986) reported eight cases of cutaneous schistosomiasis in 40 years in one hospital, and illustrated two recent cases with keloid-like lesions on the neck and upper chest, both of which contained eggs of *S. haematobium*, although the exact site of the adult worms was unclear.

Longevity

Anecdotal accounts of schistosome longevity usually describe cases of infected persons who continue to pass eggs after residence for very long periods in nonendemic areas. Usually these are incidental findings made during examination for other reasons, and in the absence of other signs or symptoms represent persistent light infections. Prolonged latent *S. japonicum* was reported by Markel et al. (1978) in a patient who shed eggs 30 years after the possibility of reinfection. Berberian et al. (1953) described a Yemeni man who had not left northern New York state for 27 years but who passed nonviable eggs of *S. haematobium* in the urine, and of *S. mansoni* in the stool. These authors cite previous publications in which viable eggs of *S. haematobium* were found in a patient after 28 years, and inviable eggs after 29 years away from endemic areas. A case of *S. haematobium* surviving for 30 years in the epididymis was described by de Gentile et al. 1988.

For *S. mansoni*, Warren et al. (1974) showed that one Yemeni man who had lived in the United States continuously for 20 years was still passing eggs. Harris et al. (1984) discussed several cases of *S. mansoni* infection in persons from Poland who had spent World War II in East Africa and who migrated to Australia in 1950. Viable eggs were found as recently as 1982 in such a man, who for at least 32 years had had no possibility of reinfection. Cook and Bryceson (1988) reported a man who had left Uganda 32 years previously, had never been out of Europe since then, and still passed 11,600 *S. mansoni* eggs in 24 hours. The same authors recorded transmission of *S. mansoni* eggs by an Indian woman who had left Uganda a mere 23 years earlier. In France, Chabasse et al. (1985) found a man who had returned permanently from Madagascar 37 years before, and whose rectal and sigmoid biopsies still contained living eggs of *S. mansoni*. At this writing, 47 years appears to be the record schistosomal Methusaleh, a case of *S. japon-*

icum infection. However, these isolated cases can give no meaningful indication of the normal life table expectation of life of these worms.

Population studies to determine the true life span of schistosomes have all been based on the finding of eggs in the excreta; however, such work can only demonstrate the period of oviposition rather than actual longevity, which may be considerably greater. Similarly, no data are available for unisexual infections, in which eggs are generally not produced. The situation is further complicated by such factors as human movements and migration, age at acquisition, intensity of infection, and the administration of antischistosomal or other drugs. Nevertheless, those studies that have been done all suggest that the mean longevity of schistosomes is much less than is suggested by the spectacular cases cited above. In a population of 218 agricultural workers of Yemeni origin in California, the mean life span of *S. mansoni* was estimated to be between 5 and 10 years (Warren et al. 1974). On St. Lucia, data were collected on a stable population of 625 persons over more than 5 years (Goddard and Jordan 1980); it was concluded that local *S. mansoni* had a mean life span of 3.3 years, with approximate 95% limits of 2.7 and 4.5 years, possibly being greater for younger hosts. A large population study on survival of *S. mansoni* in Puerto Rico (Vermund et al. 1983) found egg output to remain essentially unchanged for at least 8 years after cessation of transmission. Statistical models based upon various assumptions resulted in longevity estimates ranging from 4 to 37 years, with a mean life span probably between 5 and 15 years. In the case of *S. japonicum*, Garcia (1976) cited a report that cases persisting without reexposure for 20 years are not uncommon. For the nonhuman schistosomes, longevity is obviously related to the life span of the host (Loker 1983).

Worm Burden

Because the number of adult worms per host depends on the intensity and rate of cercarial exposure, the degree of individual resistance, and similar factors, it is unproductive to prepare an exhaustive compilation of published counts of worm number in naturally infected animals. For humans, only two examples will be cited. Careful quantitative parasitological studies at necropsy done in Brazil by Cheever (1968) revealed up to 1,608 naturally acquired *S. mansoni* worm pairs in a massively infected child, although most worm burdens were far lighter. In an Egyptian study (Cheever et al. 1977), only 22% of 59 *S. mansoni* cases had more than 100 females, and 17% of 46 *S. haematobium* cases had more than 10 females. These anecdotal reports clearly do not cover the full range of natural infections in man, and it may be assumed that the worm burden in many areas is inversely proportional to the probability of receiving treatment or of being studied scientifically. As the current review deals with the biology of the parasite, a quantitative account of infection in human populations is beyond the scope of this report.

In innumerable studies laboratory animals have been exposed to known numbers of cercariae. The proportion developing to mature adults has been termed the worm return, shown on Table 2–3 for a representative series of published reports. Here again much depends on the technique, including age of cercariae, nature and temperature of water, route of exposure, status of the host, and so forth, so that comparisons between different accounts should be made with caution.

Vertebrate Host Factors

Species

It is hardly possible to speak of schistosomes without simultaneously discussing the broad range of schistosome-definitive host relationships (Loker 1983). In nature any species of homeotherm in contact with natural waters is at risk of cercarial penetration, and if recent reports are correct even crocodilians may be susceptible to infection. Cercariae seem to be relatively nondiscriminating and at least some, or perhaps all cercariae of avian and mammalian schistosomes, will penetrate hosts of either group. The duration of subsequent parasitic life depends primarily on the species of host, ranging from minutes to hours in inappropriate hosts (where the cercariae succumb, inducing cercarial dermatitis), to decades if the luck of the draw provides the parasite with an ideal internal environment. Successful natural infection is sometimes reported with "foreign" schistosomes, as for example *S. bovis* in humans in Niger (Mouchet et al. 1988). The situation is complicated further by different host patterns of tolerance or resistance to reinfection with the same or other species of schistosomes.

Animal Models and Antischistosomal Vaccine Testing

Rodents. It is ironic that most of our knowledge of schistosome biology has been obtained from experimental work in incorrect and unnatural hosts. The widespread use of rodent models of schistosomiasis is primarily a reflection of their convenience and low cost, and not of any biological similarity to humans. The use of the mouse as a model for schistosomiasis has been reviewed by Dean (1983) and by Smithers et al. (1987).

While there are reports of *S. mansoni, S. japonicum* and other species infecting rodents in the field (Pitchford 1977, Pesigan et al. 1958), they never occur naturally in white mice or Syrian golden hamsters, and their biology in these hosts may be aberrant in significant ways. It has been shown on the island of Guadeloupe that in wild *Rattus rattus, S. mansoni* infections average 119 worms, males are 5.48 mm in mean length, and numerous eggs are produced; in wild *R. norvegicus* the average infection is 21 worms, males are 3.73 mm long, and no eggs are found in the feces (Imbert-Establet 1982).

Rodents present certain problems as models for antischistosomal vaccine development: (1) a single worm pair in a mouse affects hepatic blood flow and is said to be equivalent to a burden of about 4,000 worms in man; (2)

TABLE 2–3 Worm Return

Species/ Origin	Host/ Number	Cercarial Exposure No./Method	Worm Recovery Age/Method	Worm Return Avg. No./(%)		Reference Note
S. mansoni						
Puerto Rico	Mouse/6	150/TI	42/–	12.5		1
Puerto Rico?	Mouse/15	140/BI	63–68/P	9.31 F,		2
				24.69 M		
Puerto Rico	*Rhesus*	–/–	40–107/–		(60)	3
Puerto Rico	Mouse/280	50M/BI	7th wk/D	11.67M(23.3)		4
		50F/BI	7th wk/D	10.64F(21.3)		
	Mouse/523	100/BI	7th wk/D		(20.9)	
Puerto Rico	Mouse				(26–44)	5
Puerto Rico	Mouse/85	200/PC	20d/P		(37)	6
	Hamster/50	200/PC	40d/P		(76)	
	Rat/40	200/PC	20d/P		(15)	
Puerto Rico	Hamster/3	75/BI	v/P	48	(64)	7
	Hamster/3	150/BI	v/P	89	(60)	
Puerto Rico	Mouse/50	100/BI	4–12w/P	22.9	(22.9)	8
	Mouse/50	100/IP	4–12w/P	21.4	(21.4)	
	Hamster/23	160/IP	4–12w/P	28.7	(17.9)	
	Hamster/22	160/PCS	4–12w/P	46.6	(29.2)	
	Hamster/25	160/PCU	4–12w/P	29.0	(18.1)	
	Rat/27	100–1,000/IP	5–14w/P	3.0	(0.9)	
	Rat/12	250–1,000/ PCS	6–12 + w/P	33.	(2.3)	
	Guinea Pig/4	1000/IP	6–12w/P	4.8	(0.5)	
	Guinea Pig/11	500–2,000	4–12w/P	280	(22.9)	
	Rabbit/12	500–15,000/ IP	6–9w/P	14.9	(0.3)	

Location	Species/No.	Dose/Route	Duration	Value	(%)	Ref.
	Rabbit/12	500–15,000/PCE	6–9w/P	700	(11.8)	
Puerto Rico	Rhesus/6	750/PC	– /P	476	(63.5)	9
Puerto Rico	Mouse/33	200–400/TI	49–52d/P+D	39.3 M / 56.1 F		10
Puerto Rico	Mouse/8CFW	200/TI	49d/D	52.0	(26.1)	11
	Mouse/9RAP	200/TI	49d/D	51.7	(25.9)	
Puerto Rico	Mouse/55	50M+50F/TI	–/D	19M+17F		12
	Mouse/55	50M/TI	–/D	17		
	Mouse/55	50F/TI	–/D	15		
	Mouse/20	50M+50F/TI	–/D	19M+17F		
	Mouse/20	50M/TI	–/D	16		
	Mouse/20	50F/TI	–/D	15		
Puerto Rico	Hamster/15	50/BI	4–16w/D	17.4	(34.3)	13
	Hamster/28	100/BI	4–16w/D		(32.9)	
	Hamster/30	200/BI	4–16w/D		(30.4)	
	Hamster/21	500/BI	4–16w/D		(32.1)	
	Mouse/184	100/BI	4–16w/D	23.2	(23.2)	
	Mouse/70	200/BI	4–16w/D		(20.9)	
	Mouse/55	500/BI	4–16w/D		(22.3)	
	Cotton Rat/23	100/BI	4–16w/D		(20.2)	
	Cotton Rat/12	200/BI	4–16w/D		(17.0)	
	Cotton Rat/12	500/BI	4–16w/D		(15.2)	
	Guinea Pig/36	100/BI	4–16w/D		(4.7)	
	Guinea Pig/50	200/BI	4–16w/D		(6.6)	
	Guinea Pig/36	500/BI	4–16w/D		(6.6)	
	White Rat/87	100/BI	4–16w/D		(3.8)	
	White Rat/86	200/BI	4–16w/D		(3.1)	
	White Rat/77	400/BI	4–16w/D		(2.2)	
	Cat/4	Many			(10–45)	
	Rabbit/6	300–23,500			(0–18)	
	Dog/8				(0.0)	

TABLE 2–3 Worm Return (continued)

Species/ Origin	Host/ Number	Cercarial Exposure No./Method	Worm Recovery Age/Method	Worm Return Avg. No./(%)	Reference Note
S. mansoni					
Puerto Rico	Mouse/30	40/PCS	4–16w/P	(38.4)	14
	Hamster/25	40/PCS	4–16w/P	(53.8)	
	Rabbit/23	40/PCS	4–16w/P	(23.2)	
	White Rat/25	40/PCS	4–16w/P	(10.3)	
	Guinea Pig	40/PCS	4–16w/P	(22.7)	
Puerto Rico	Rabbit/1	10,000/IP	42d/D	60 (0.6)	15
	Rabbit/1	10,000/EI	42d/D	3,827 (38.3)	
	Hamster	600/IP	42d/D	145 (24.2)	
	Hamster	1,000/PCS	42d/D	512 (51.2)	
Puerto Rico	Mouse/35	35/TI	8w/P	(36.6)	
Brazil	Mouse/38	40/TI	8w/P	(27.2)	
Egypt	Mouse/40	40/TI	8w/P	(32.5)	
Tanzania	Mouse/38	40/TI	8w/P	(29.9)	
Puerto Rico	Mouse/10	78/PCS	6w/P	(39.2)	16
	Rat/10	500/PCS	4w/P	(11.4)	
	Mastomys/10	100/PCS	6w/P	(45.4)	
	Macaca/7	1,000 PCS	6–7w/P	(49.1)	
Liberia	Guinea Pig/–	850–5,000	2–6m	(1–2)	17
Tanzania	Rhesus/6	750/PC	–	344 (45.9)	18
Brazil (LE)	Mouse/–	80–100/TI	8w	(about 20)	19
Puerto Rico	Mouse/18	50 M or F/BI	35d/P	(15.9M, 8.8F)	20

70

	Host/No.	Dose	Day/Route			Ref.
S. mansoni						
Egypt	Mouse/600	90/BI	70–77d/D	18.5	(20)	21
Tanzania	Mouse/20	50/BI	60d/P	2.1	(4.2)	22
	Mouse/20	100/BI	60d/P	3.1	(3.1)	
	Mouse/20	200/BI	60d/P	9.9	(5.0)	
	Mouse/20	300/BI	60d/P	12.5	(4.2)	
	Mouse/20	400/BI	60d/P	15.3	(3.8)	
	Mouse/20	500/BI	60d/P	24.6	(4.9)	
	Hamster/6	50/BI	60d/P	1.6	(3.2)	
	Hamster/6	100/BI	60d/P	6.1	(6.1)	
	Hamster/6	200/BI	60d/P	17.8	(8.9)	
	Hamster/6	300/BI	60d/P	21.2	(7.1)	
	Hamster/6	400/BI	60d/P	53.4	(13.3)	
	Hamster/6	500/BI	60d/P	68.9	(13.8)	
S. japonicum						
— China	Guinea Pig/84	60–175/PC	–/D		(27.7)	23
	Mouse/20	50M + 50F/ PC?	41d/P	43.2		24
Taiwan	Mouse/20	50M + 50F/ PC?	41d/P	66.7		
Japan	Mouse/26	50M + 50F/ PC?	34d/P	85.8		
Philippines	Mouse/20	50M + 50F/ PC?	35d/P	82.7		
Taiwan?	Hamster/5	75–80/PC	–/PD	35.8	(46.4)	25
China	Mouse/7	100/PCS	2–4w/P	46	(46)	26
	Mouse/6	25/PCS	8w/P		(43)	
	Mouse/6	50/PCS	8w/P		(45.6)	
	Mouse/6	100/PCS	8w/P		(46.8)	
Philippines	Mouse/–	84 total/PCS	22–46d/D	61	(72.6)	27
S. haematobium						
Sudan	Hamster/54	200/PCS	170d/P	24.8	(12.4)	28
	Hamster/62	200/PCS	250d/P	24.2	(12.1)	

TABLE 2–3 Worm Return (*continued*)

Species/Origin	Host/Number	Cercarial Exposure No./Method	Worm Recovery Age/Method	Worm Return Avg. No./(%)		Reference Note
S. haematobium						
S. Africa	Hamster/54	200/PCS	170d/P	26.2	(13.1)	
	Hamster/63	200/PCS	250d/P	24.6	(12.3)	
Egypt	Mouse/47	150/PC	6–12w/P	5.3	(3.5)	29
	Albino Rat/5	500/PC	6–14w/P		(0.12)	
	Guinea Pig/14	500/PC	6–14w/P		(0.16)	
	Hamster/65	200/PC	6–20w/P	33.8	(16.9)	
	Rabbit	1,500–11,000/PC	7–23w/P	0.0	(0.0)	
Tanzania	Hamster/6	40/PC	90d/P	3.4	(8.5)	30
	Hamster/12	75/PC	90d/P	5.3	(7.1)	
	Hamster/6	150/PC	90d/P	12.5	(8.3)	
	Hamster/6	235/PC	90d/P	17.5	(9.4)	
	Hamster/6	250/PC	90d/P	15.2	(6.1)	
	Hamster/6	350/PC	90d/P	26.9	(7.6)	
	Hamster/6	400/PC	90d/P	38.8	(9.7)	
	Hamster/6	550/PC	90d/P	48.4	(8.8)	
	Hamster/6	650/PC	90d/P	31.2	(4.8)	
	Hamster/6	1,000/PC	90d/P	91.3	(9/1)	
Iran	Hamster/55	200/PCS	75–170d/P	49	(24.7)	31
Nigeria	Hamster/55	200/PCS	75–170d/P	37	(18.5)	
S. Africa	Hamster/61	200/BI	75–170d/P	42	(21.)	32
Ghana K strain	Hamster/148	200/BI	50–d/P	70.3	(35.2)	33
Ghana P strain	Hamster/156	200/BI	50–d/P	72.7	(36.4)	
Iran	Hamster/149	200/BI	50–d/P	35.5	(17.8)	
Mauritius	Hamster/132	200/BI	50–d/P	41.4	(20.7)	

	Host/No.	Dose	Time	No.	(%)	Ref.
S. intercalatum						
Cameroon	Mouse/52	120/TI	40–80d/P	23.4	(20)	34
Zaire	Mouse/53	120/TI	40–80d/P	24.7	(21)	35
Cameroon	Mouse/–	100–600M/BI	60d/D		(35.0)	
	Mouse/–	100–600F/BI	50d/D		(12.2)	
Zaire	Mouse/–	100–600M/BI	60d/D		(39.8)	
	Mouse/–	100–600F/BI	60d/D		(18.7)	
Cameroon	Hamster/50	200/–	—		(17.3)	36
S. rodhaini						
Uganda	Rabbit/–	–/BI	—		(10–15)	37
	Hamster/–	–/BI	—		(43.2)	
	Mouse/–	–/BI/TI	—		(39.1)	
Orientobilharzia turkestanicum						
Iran	Calf/8	8,000/LI	9–18w/–	2,647	(37.6)	38
	Sheep/5	5,000/LI	9–18w/–	1,694	(33.9)	
	Goat/2	5,000/LI	9w/–	1,125	(22.5)	
	Buffalo Calf/1	8,000/LI	9w/–	773	(9.6)	
	Wild Boar/1	5,000/LI	11w/–	26	(0.6)	
	Albino Mouse/15	50–400/BI	—	80	(26.9)	
	Hamster/11	50–200/BI	—	42	(34.2)	
	Mastomys/10	100–400/BI	—	62	(25.8)	
	Rabbit/6	200–1,500/–	—	157	(21.0)	
	Tatera/10	200–1,000/–	—	90	(15.1)	
	Nesokia/7	200–1,000/–	—	98	(16.4)	
	Rattus/6	200–1,000/–	—	24	(3.7)	
	Mouse/8	100–440/–	—	5	(1.8)	

TABLE 2–3 Worm Return (continued)

Species/ Origin	Host/ Number	Cercarial Exposure No./Method	Worm Recovery Age/Method	Worm Return Avg. No./(%)	Reference Note
S. mattheei					
S. Africa	Sheep/45	–/LI	45–146d/P	2,202 (22.2)	39
S. Africa	Sheep/12	100M/LI	74–76d/–	50.8	40
	Sheep/12	100F/LI	74–76d/–	66.7	
S. Africa	Hamster/34	–/BI	—	(20.8)	41
S. margarebowiei					
Botswana	Hamster/24	100/BI	28–60d/P	39.8	42
Schistotomatium douthitti					
Michigan	*Peromyscus*/36	54–400	14–383d/–	(53.4)	43

Abbreviations: Exposure method as stated in reference—BI, body immersion; EI, ear immersion; IP, intraperitoneal injection; PC, percutaneous; PCE, percutaneous on ear; PCS, percutaneous with skin shaved; PCU, percutaneous with skin unshaved; TI, tail immersion. Worm recovery as stated—d, days; w, weeks; D, dissection; P, perfusion; V, variable.

Reference notes: 1, Bennett and Gianutsos 1978; 2, Berg 1953; 3, Cheever and Powers 1969; 4, Evans and Stirewalt 1957; 5, Cohen and Eveland 1984; 6, Miller and Wilson 1980; 7, Moore and Sandground 1956; 8, Moore et al. 1949; 9, Nelson and Saoud 1968; 10, Robinson 1959; 11, Robinson 1957b; 12, Stirewalt and Fregeau 1968; 13, Stirewalt et al. 1951; 14, Warren and Peters 1967; 15, Yolles et al. 1949; 16, Smithers and Terry 1965; 17, Lagrange and Scheecqmans 1949; 18, Nelson and Saoud 1968; 19, Pellegrino and Faria 1965; 20, Liberatos 1987; 21, Standen 1949; 22, Purnell 1966; 23, Bang and Hairston 1946; 24, Hsü and Hsü 1960; 25, Moore and Sandground 1956; 26, Moloney and Webbe 1983; 27, Pesigan et al. 1958; 28, James and Webbe 1975; 29, Moore and Meleney 1954; 30, Purnell 1966; 31, Webbe and James 1971b; 32, Wright and Bennett 1967a; 33, Wright and Knowles 1972; 34, Bjorneboe and Frandsen 1979; 35, Frandsen 1977; 36, Wright et al. 1972; 37, Fripp 1968; 38, Massoud 1973a; 39, McCully and Kruger 1969; 40, van Rensburg and van Wyk 1981; 41, Wright et al. 1972; 42, Southgate and Knowles 1977; 43, Short 1952a.

the average life span of the schistosomes is longer than that of mice, thereby eliminating long-term studies; (3) the metabolism of rodents leads to variant actions of certain drugs; for example, antimonials are effective in man but not in the mouse; teroxalene is good in the mouse but not tolerated in humans (Sturrock 1986); (4) the immunologic responses of mice may be only marginally relevant to the development of antischistosomal vaccines intended for man.

Hamsters (*Mesocricetus auratus*) have been used widely and are the hosts of choice for *S. haematobium*. Other rodents and lagomorphs each present specific difficulties for vaccine development and testing.

Primates. (Sturrock 1986, Sturrock et al. 1988). Monkeys used for *S. mansoni* include rhesus (*Macaca mulatta*), capuchin (*Cebus apella*), baboon (*Papio spp.*), and various cercopithecines (*Cercopithecus aethiops*). Species susceptible to *S. haematobium* and *S. japonicum* include: vervet, rhesus, capuchin, taloipin, and baboon.

Like rodents, primates vary greatly in their response to exposure to schistosomes. Hepatic granuloma formation differs among species of primates, and the chimpanzee is the only one that develops a Symmers-like pipestem fibrosis. Although pathologic studies on chimpanzees were conducted in decades past, their employment at present is essentially a nonissue. The cost and rarity of some other species, as well as animal rights issues, demand stringent justification for the use of any primates, in schistosome vaccine trials.

Rhesus monkeys may demonstrate partial or complete self-cure and solid natural resistance to schistosomal challenge, making them inappropriate for certain vaccine studies. Cheever et al. (1988) found that capuchin monkeys became resistant to reinfection with *S. haematobium* 1 year after an initial infection of 500 cercariae, and completely refractory when challenged 2–5 years after an initial exposure of 1,000 to 2,000 cercariae.

Baboons may present an adequate compromise primate host for vaccine development. Adult *S. mansoni* worms are found in the same site in baboons and humans. Both develop partial resistance, possibly related to intensity and duration of primary infection and size and frequency of challenge. However, baboons show little or no fibrosis, although there is some increase in liver collagen level. They do not develop collateral circulation, there is no correlation between infection and spleen weight; and they do not develop marked hepato- or hepatosplenomegaly. Baboons develop a natural suppression of egg production in *S. mansoni* and are more resistant to *S. haematobium*, which in this species lives mostly in the intestinal veins.

Host Permissiveness

The idea of "permissive" and "nonpermissive" mammalian hosts for schistosomes was introduced by Cioli et al. (1977), and these terms have since gained widespread acceptance. For *S. mansoni,* Knopf (1982) has reviewed and listed the mammalian species in each category: hamsters and mice are in general found to be good hosts; rats, guinea pigs, rabbits, cats, dogs, and various other mammals, poor hosts.

Work using different breeds or strains of mammalian hosts is legion; indeed, it is hardly possible to compare publications that refer only to "mice" "sheep," and so forth. In a recent study, Jones and Kusel (1985) have compared the inheritance of responses to a single pool of Puerto Rican S. mansoni infection in two pairs of inbred mouse strains. They found that even when the worm burden was similar, some mouse strains had higher fecal egg output and larger numbers of eggs in their tissues, and differed in the degree of splenomegaly and pattern of antibody response as compared to other strains.

Available data suggest a complex interplay not only between species of hosts and parasites, but between particular genetic stocks of each; for example, Dean (1983) has summarized information about resistance to schistosomes in various inbred strains of mice. More recent evidence suggests that genetically determined structural features of the portal system are correlated with relative resistance to schistosome infection in 129/J and 129/Ola mice (Garcia et al. 1983, Mitchell 1989, Elsaghier et al. 1989, Elsaghier and McLaren 1989). Some genetic analysis has been carried out by crossing, suggesting dominance of susceptibility to S. mansoni and S. japonicum and control by one or two genes (Wright et al. 1988).

The immune response is itself under genetic control and loci controlling protective immune responses have been identified in mice (Correa-Oliveira et al. 1986, 1988). The murine H-2 complex influences the response to schistosomal infection, and to specific antigens (Jones et al. 1983, Kee et al. 1986). On the other hand, Fanning and Kazura (1984) showed that various inbred mouse strains developed high, medium or low worm burdens in response to identical exposures but concluded that parasite number was controlled polygenically and not related to the major histocompatibility complex in these mice. The degree of hepatic fibrosis and granuloma formation in mice differs among genetic strains (Cheever et al. 1983, 1987). A mouse strain, P/N, has been found that fails to respond to antischistosomal immunization.

In humans the immune response to schistosomiasis is known to vary among different groups of patients (Gazzinelli et al. 1987). The HLA system has been correlated with responses to S. japonicum (Ohta et al. 1987, Wang et al. 1984, Watt and Sy 1988); S. haematobium (Wishahi et al. 1989); and S. mansoni (Abdel-Salam et al. 1986, El-Hawy et al. 1989). The immunological literature should be consulted for more specific information.

The generally poor development of S. mansoni in guinea pigs has often been noted (Moore et al. 1949, Kuntz 1961a, Lagrange and Scheecqmans 1949, Stirewalt et al. 1951, Warren and Peters 1967) and the tendency of some geographic isolates to become hermaphroditic in these hosts is discussed elsewhere. LoVerde et al. (1985b) have suggested that a cercarial pool may be altered by repeated passage through one or another species of host—in the case reported, baboons and mice—so that worms recovered from the two hosts show some divergent characters. Thus, although most investigators have studied the effects of schistosomes on their hosts, it may be equally rewarding to look into the reverse situation.

Schistosoma rodhaini is an African species closely related to *S. mansoni*. Its principal natural hosts are rodents, and mice and hamsters were good hosts. However, rabbits and guinea pigs were poor. *S. rodhaini* will not infect primates, including man (Fripp 1968).

Surgical Transfer of Worms. *Schistosoma mansoni* worms have been transferred surgically between hosts in an attempt to clarify host-parasite relationships. Michaels (1970) transferred schistosomula and worms between hamsters and guinea pigs in all combinations. It was found that for homologous transfers egg passage was as expected: hamsters receiving worms from hamsters passed eggs, and guinea pigs receiving worms from guinea pigs did not. All hamsters receiving worms donated by guinea pigs passed *S. mansoni* eggs after a brief interval, indicating that the guinea pig does not permanently inhibit development. In the reverse direction, prior residence in the hamster increased the worms' reproductive capacity when transferred to guinea pigs, but only when the transferred worms were sexually immature. The reason for the reproductive inhibition of transplanted adult worms was not clear. In a later study, Cioli et al. (1977) used rats, mice, and hamsters as donors and recipients. *S. mansoni* worms transferred from rats to hamsters were found to mature rapidly and normally, but adult worms from mice implanted into rats deteriorated rapidly. When such degenerating worms were retrieved and implanted into mice, they recovered and resumed egg production. Worms grown from cercariae in vitro behaved as if they were in a nonpermissive host, and only inviable eggs were produced (Basch 1981a, b). Implantation into mammals of cultured schistosomula and juvenile worms yielded normally ovigerous worm pairs, but when worms had already paired in vitro, full maturity with viable eggs could not be achieved (Basch and Humbert 1981). In a subsequent study (Basch and Rhine 1983), some of the cultured male and female adults did become mature and fertile when coimplanted with partners derived from mouse infections.

For *S. japonicum* the list of permissive and nonpermissive hosts is quite different. Earlier work was reviewed by Kuntz (1955). In extensive studies Ito (1954) showed that for worms of Japanese origin, rabbits showed the highest worm return and survival. In mice, worms matured more slowly and disappeared more rapidly. In guinea pigs the infection rate was lower and worms matured still more slowly; as with *S. mansoni*, very few eggs were found in the feces of this host. The worst rodent host for *S. japonicum* was the rat (presumably *R. norvegicus*), in which the infection rate was poor and most worms died rapidly. In natural *S. japonicum* infections in the Philippines rats had by far the lowest daily egg output of 7 host species studied. Hatchability of eggs from natural infections ranged from 10.6% in rats to 71.9% in cattle; 42.4% of eggs from humans hatched to produce miracidia.

On Taiwan Chiu and Kao (1973) looked at the prepatent period of *S. japonicum* in 16 species of mammals. Suitable hosts, on the basis of a prepatent period of less than 55 days, were: mouse (Swiss Webster), Syrian golden hamster, domestic rabbit, cat, dog, goat, pig, calf, bandicoot, and squirrel. Unsuitable hosts were: laboratory rat (Long Evans), and brown rat (*R. norvegicus*), spinous country rat (*R. coxinga*), guinea pig, mongoose, and *Macaca cyclopis*. It was found by Hsü and Hsü (1958a) that the dimensions of eggs of *S. japonicum* varied with their geographic origin and laboratory host.

For *S. haematobium,* Gordon et al. (1934) found guinea pigs very unsuitable. A total of three immature worms were recovered from four animals exposed to from 200 to 4,600 cercariae. Moore and Meleney (1954) infected mice, rats, hamsters, guinea pigs, rats, and rabbits with cercariae of an Egyptian isolate, finding some eggs in feces of mice and hamsters. The latter were considered the most suitable laboratory host. Ten mammalian species, including rodents, bandicoots, hedgehogs, and primates, were tested by Kuntz (1961b), with generally poor results. With respect to other species of schistosomes, only scattered reports are available concerning their development in various mammalian hosts. It was found by Brumpt (1936) that *S. bovis* would not develop completely in two species of *Macaca* and that worms were stunted and infertile in rats. Lengy (1962b) found that 6 months post-infection the mean length of male *S. bovis* was 17 mm in sheep but only 10.2 mm in guinea pigs, with intermediate sizes in mice, hamsters, and rats. In India, Liston and Soparkar (1918) exposed various mammals to cercariae of *S. spindale*. A rhesus monkey showed no signs of infection despite exposure to 100,000 cercariae. In guinea pigs only male worms developed but a goat, as expected, developed an overwhelming and lethal bisexual infection after exposure to water "swarming with cercariae." With the same schistosome Fairley et al. (1930) also found good development in goats, although guinea pigs exposed to 42,000 to 72,000 cercariae developed only few immature pairs lacking eggs.

Experimental infection of various mammals has been reported, *inter alia*, for *S. intercalatum* by Wright et al. (1972); *S. incognitum* by Sinha and Srivastava (1956); *Orientobilharzia turkestanicum* by Machattie (1936); *Schistosomatium douthitti* by Price (1931); and *Heterobilharzia americana* by Goff and Ronald (1981).

Host Immunity and Parasite Regulation

The interplay between the immunologic armamentarium of mammalian hosts, and the defense mechanisms of schistosomes, is a subject of utmost importance to schistosome biology, particularly in view of current efforts to develop antischistosomal vaccines. The extended literature on this subject, most of it from experimental rodent hosts, has been reviewed repeatedly (e.g., by McLaren and Smithers 1987; Vignali et al. 1989; Sher et al. 1989) and is simply too large and diverse to be managed in this discussion of schistosome biology.

Host Age and Sex

It is generally agreed among authors that younger animals are more easily infected than are older ones. The effect of host sex and age on development of *S. mansoni* in mice and hamsters, and of *S. haematobium* in hamsters was studied by Purnell (1966). With *S. mansoni,* the percentage of worm recovery declined rapidly up to one month of host age and was stable thereafter; with *S. haematobium* no such relationship was demonstrated. With both parasites male hosts were more favorable, especially so in the case of male hamsters and male *S. haematobium*. The authors concluded that sex of the host may influence the sexual balance of its schistosome population. Another study of host sex in relation to development of *S. mansoni* was that of Rombert and Ordens (1978), who exposed 353 mice each to 250 male, female, or mixed cercariae by tail immersion. Mean mixed worm recovery was almost identical in male and female hosts. Female mice had slightly fewer unisexual worms of each sex than did males. Although Rombert did not apply statistical methods in his study, the differences in worm numbers between the sexes did not appear significant. Regarding unisexual female worms alone, Shaw and Erasmus (1981) found no differences in development in male or female mouse hosts.

Schistosoma haematobium from Iran, Mauritius, and Ghana demonstrated slightly higher infection rates in female than in male hamsters (24% vs. 20% worm return), with a sex ratio (M:F) in male hosts of 1.53:1 and in female hosts of 1.74:1 (Wright and Knowles 1972). In *S. japonicum,* Li (1958) found that unisexual male worms developed much faster in male than in female mice, whereas unisexual female worms became "only slightly" larger in female than in male mice.

Host Hormonal Balance

Various authors have attempted to manipulate host hormonal balance to alter subsequent schistosomal infection. Berg (1953) castrated 5- to 6-week-old male mice and then infected them with 140 cercariae of *S. mansoni*. He reported that the castrated mice yielded fewer male worms than intact controls, a difference significant at the 5% level. The same author (Berg 1957) found that testosterone treatment of both castrated and intact mice reduced worm burdens. These results were contradicted by Robinson (1959), who found no predictable relationship of castration and testosterone treatment to survival of worms. Robinson (1957b) used both CFW and RAP strain mice, castrated or intact, finding no difference in the numbers or reproductive development of male or female worms among the four groups. Working with hamsters, Combescot et al. (1971, 1974) found that ovariectomy of female hamsters did not affect the development of *S. mansoni,* but that implantation of a 20-mg estradiol pellet into the operated animals inhibited or prevented growth of the schistosomes. In a similar study Barrabes et al. (1979) found about half the expected number of *S. mansoni* adults when "castrated" female hamsters were treated with estradiol or testosterone, but approximately the same as in control mice when progesterone was administered.

Steroid metabolism (Briggs 1972) appears to be different in infected and uninfected mice. Lagrange and Scheecqmans (1949) found that uninfected mice accepted 500 micrograms of WIN1303 with no problem, while mice infected with *S. mansoni* died with a dose of 75 to 100 ug. Repeated small doses of certain steroids stopped egg output, caused sterilization and deterioration of worms, and killed eggs. Some 30 estrogen and androgen steroids were all negative for antischistosomal effect; the only active compounds consisted of 21 carbons with a ketone on C20 and an alcohol or ester on C21. In baboons infected with *S. mansoni* or *S. japonicum,* Newsome (1963) found that dexamethasone greatly reduced egg output, while C-nor-prednisone acetate was without effect at the concentrations employed. The natural hormonal changes of pregnancy were studied by Preston and El-Khatib (1977): no difference was detected in the number of worms developing in mice exposed when pregnant, those becoming pregnant after exposure, or in control female mice.

The role of host hormones in controlling survival and development of *S. mansoni* was reviewed by Knopf (1982), who stated that "Sexual maturation in mammalian hosts of schistosomes is a hormone-dependent developmental process." Hypophysectomy of rats was shown to delay elimination of juvenile worms, prolong survival of males, increase the mean length of adults and vitelline cell staining in females, and improve the number and condition of eggs so that some contained hatchable miracidia. In mice rendered hypothyroid, worms were smaller and matured more slowly; in mice made hyperthyroid, worms were larger and matured more rapidly, but in both cases the total egg burden was greater than in euthyroid controls. Rats are normally nonpermissive hosts. In thyroidectomized rats the worms could complete their life cycle, and host livers contained more than 10 times more eggs than the livers of intact rats. Small numbers of infective miracidia hatched from these eggs, whereas there was no hatch from livers of intact rats (Knopf and Linden 1985).

A variety of mammalian regulatory peptides has been identified immunocytochemically in schistosomes (Basch and Gupta 1988; Gupta and Basch 1989), but whereas transferrin and certain growth factors were stimulatory in vitro, an extensive study failed to demonstrate the previously claimed effect of insulin on *S. mansoni* (Clemens and Basch 1989a, b).

Host Nutrition

Early studies on host nutritional status and schistosome development were made by Krakower et al. (1940), working with *S. mansoni* in Puerto Rico. These authors found that the parasite survived somewhat better in vitamin A-deficient than in normal rats. Subsequently (Krakower et al. 1944) they fed some guinea pigs on a vitamin C-deficient diet, some on the same with the vitamin C replaced, and some on grass. Deficient eggshells and disintegrated eggs were found in many of the worms in scorbutic animals or those with insufficient vitamin C replacement. The reproductive system of female worms appeared normal in all, but some in deficient animals had a granular mass in the uterus, never seen in worms in the normal animals. In a better host, the mouse, multiple nutritional deficiencies were induced in animals fed on a Torula yeast diet deficient in Factor 3 (selenium), vitamin E, and cystine, resulting in multiple dietary necrotic degeneration affecting the heart, liver, muscle, and kidneys (DeWitt 1957a, b). In such hosts the schistosomes failed to attain normal size and with very few exceptions did not become sexually mature. In female worms the posterior part of the body, containing the vitelline glands, was most severely reduced. No eggs were produced, and the few found in the uteri of female worms were pigmented and granular. The number of worms harbored by mice on the deficient diet was 69% more than in controls, demonstrating a diminution of host resistance. Males of an Egyptian isolate of S. mansoni were smaller than normal and had aberrant reproductive systems when grown in mice fed a protein-deficient diet. Egg production was reduced in low-protein fed mice as compared with normally nourished controls (De Meillon and Paterson 1958).

Some subsequent studies on host nutrition and schistosomiasis were reviewed by Knauft and Warren (1969), who placed white mice on isocaloric diets containing 4%, 8%, 12%, or 20% protein, or on 25% and 50% calorie-deficient diets. Some mice exhibiting diet-related effects were exposed, with controls, to cercariae of a Puerto Rican S. mansoni. While total worm burden was essentially unaffected, all the deficient diets resulted in a marked decrease in the number of eggs in the liver per worm pair, and in animals on the poorest protein diet the liver granuloma size was greatly reduced. There was a slight but significant reduction in the size of both male and female worms in mice fed the most protein- and calorie-deficient diets. Severe calorie deficiency appeared to affect the parasite more than the host.

Also in S. mansoni-infected mice, Cornford et al. (1983) have shown a rapid decline in the glycogen content of male schistosomes in fasted hosts, demonstrating that the glycogen reserves of male worms are more labile than those of females. A similar phenomenon was observed in S. japonicum and S. haematobium.

A different host-parasite system, that of S. mattheei in calves, was investigated by Lawrence (1973). During the period of early peak egg output, 8 to 15 weeks post infection, a significant correlation was observed between daily fecal egg output per parasite and the plane of nutrition at low infection rates. As parasite numbers increased, there was a significant reduction in egg output in animals on a high, but not in animals on a low, plane of nutrition.

In humans, DeWitt et al. (1964) studied the effects of improving the nutrition of malnourished people infected with S. mansoni in Puerto Rico. Initial nutritional inadequacies were primarily of protein, ascorbic acid, riboflavin, niacin, and calories. While general benefits accrued to persons receiving dietary supplementation for 9 months, no clear effect was seen upon schistosome egg counts in comparison with unsupplemented controls.

Other Host Factors

In an unusual investigation, Carmichael and Muchlinski (1980) followed the effects of hibernation of Zapus hudsonius (meadow jumping mice) on *Schistosomatium douthitti*. Schistosome infections were not different in nonhibernating control mice or in mice hibernated for 100 or 150 days. In both groups worm count and degree of maturity of eggs in female worms were comparable.

3

Adult Schistosomes and Eggs

The male-female separation described in Chapter 1 has led to numerous consequences for schistosome biology. Many authors have reported differences, other than the obvious morphological and functional ones, between mature male and female schistosomes. Summaries of data on this topic were prepared by Coles (1973, 1984). Table 3–1 lists additional points of difference between male and female worms.

Histological sections of adult male and female *S. haematobium* were shown in a series of beautiful and accurate lithographs by Looss (1895) (Fig. 3–1), which still serve as an excellent guide to adult worm morphology and histology.

The Male Schistosome

The typical configuration of the male schistosome of the species infecting man is familiar to parasitologists, but this form represents only one of many variations found in the family. Illustrations of many different schistosomes of birds and mammals are found in the works of Skrjabin (1951) and Yamaguti (1958, 1971), which should be consulted for a revealing iconography of schistosomal morphology. Figure 1–11 shows a small selection of the variety of appearances of male schistosomes, flattened or cylindrical, with expanded or reduced gynecophoral canals of various sizes (Table 3–2). Females are generally long and cylindrical but in some cases, as in the genus *Dendritobilharzia* (Ulmer and vandeVusse 1970), they are broad and leaflike and resemble the male in body shape. It is in the avian schistosomes, whose biology is virtually unexplored, that the greatest differences from the "conventional" schistosome shape are found, and investigations of these forms should be encouraged as a likely source of new information.

TABLE 3–1 Differences Between Male and Female Schistosomes

Property	Observation	Reference Note
	Normal Physiology	
S. mansoni		
Amino acids	M uses more Glu, Asp, Ala; F more	1
	Tyr	2
	Uptake of Arg, Lys, His, Phe, Ser,	
	Ornithine higher in F	3
	Minor differences between sexes in	
	Met	4
	Proline uptake much higher in M	5
	F transports more glycine	
Polypeptides	M produces 5 not made by F	1
	F produces 4 not made by M	7
	F band at 29 kd not found in M	8
Proteins	Gender-specific	9
Glycoproteins	Gender-specific	10
Enzymes	Consult for review	11
	F has higher proteolytic activity	12
	F hemoglobinase 3–5× more active	13
	than M	
	Acid phosphatase uniform in F; high in	14
	tegument of M	
	Phosphoglucose isomerase pattern	15
	differs	
	M digests gelatin fibers; F does not	16
Actin	More RNA and protein in M than in F	17
Lipids	M has 22.97 mg/g; F has 55.13 mg/g	18
	wet weight	
Cholesterol	Much more in M than in F	19
	Reserves in M more labile than F when	20
	fasted	
Sugars	F transports more glucose and 2-	21
	deoxyglucose	
	Glucose uptake higher in F	22
Ingestion	F ingests 330,000; M 39,000 red cells/hr	23
Metabolism	Uridine incorporation 13–16× more in	
	M;	
	Thymidine incorporation 4× more in F	24
Physiology	Surface electrophysiology differences	
	F more sensitive to Ca and Mg ions	25
Longevity	M hardier and live longer than F	26
S. japonicum		
Amino acids	M uses more Glu, Asp, His; F more	27
	Tyr, Pro	
Enzymes	F acid protease has higher activity than	28
	M	
Metabolism	F more active	29
S. haematobium		
Metabolism	F uses more O_2 and has higher	30
	cytochrome activity	

84

TABLE 3–1 (continued)

Property	Observation	Reference Note
	Drug Effects	
S. mansoni		
Astiban	Concentrated by F 3–5× more than M	31
Hycanthone	M more sensitive than F	32
Oxamniquine	M more sensitive than F	33
	Increased M triglycerides	
	Affects M lipid pattern more than F	
Niridazole	F more sensitive than M	34
Biogenic amines	Different effects in M and F	35
Tartar Emetic	F take up more Sb than M	36
Praziquantel	Contractures less in F	37
S. haematobium		
Niridazole	F more sensitive than M	38

Reference notes: 1, Bruce et al. 1972; 2, Cornford and Oldendorf 1979; 3, Chappell 1974; 4, Senft 1968; 5, Uglem and Read 1975; 6, Atkinson and Atkinson 1982; 7, Atkinson and Atkinson 1980a; 8, Ruppel and Cioli 1977; 9, Rotmans and Burgers 1984; 10, Norden and Strand 1984; 11, deBoissezon and Jelnes 1982; 12, Grant and Senft 1971; 13, Timms and Bueding 1959, Cesari et al. 1981; 14, Cesari 1974; 15, Jelnes 1983; 16, Cesari et al. 1981; 17, Davis et al. 1985; 18, Khayyal et al. 1980; 19, Fried et al. 1983; 20, Cornford et al. 1983; 21, Uglem and Read 1975; 22, Cornford and Huot 1981; 23, Lawrence 1973; 24, Pica Mattoccia et al. 1982; 25, Siefker et al. 1983; 26, Mayer and Pifano 1942; 27, Bruce et al. 1972; 28, Yang and Chow 1981; 29, Yang and Chow 1981; 30, Moczon and Swiderski 1983; 31, Browne and Shulert 1963; 32, Cioli and Knopf 1980; Lee 1972, Pica Mattoccia et al. 1982; 33, Foster and Cheetham 73, Popiel and Erasmus 1982, Khayyal et al. 1980, Davies and Jackson 1970; 34, Hess et al. 1966, Striebel 1969, Striebel and Kradolfer 1966; 35, Tomosky et al. 1974; 36, Molokhia and Smith 1968; 37, Siefker et al. 1983; 38, Moczon and Swiderski 1983.

Comparative studies of schistosome growth and morphology are generally unrewarding without considering the sources of variation, both biological and technical, in reported dimensions and contour. Biological variations arise from the genetic makeup of the schistosome and its hosts, the number of worms present, age of the infection, and similar factors. Technical variations come from differences in handling and techniques, such as: obtaining the worms by perfusion or by dissection; the method and extent of relaxation, if any; means of fixation; degree of compression; standard procedure for measurement; and so forth. Such methodologic information is frequently lacking from published accounts of schistosome morphology and dimensions, relegating comparisons almost to guesswork. Therefore no efforts at completeness have been made in compiling Table 3–1, lest the impression be given that the reported measurements of the worms are of more than very general utility.

The Male Reproductive System

General Description. The male reproductive system is usually far simpler than that of the female, and is relatively uniform in the commonly considered *Schistosoma* infecting man and domesticated animals (Fig. 3–2). The system was described in detail by early authors such as Looss (1895), but prior to

Figure 3–1 Lithographs of morphology and histology of *S. haematobium,* made at the end of the nineteenth century. Figures 1 to 18, Males. 1, Four worms, to show the variations in gut shape; 2, Sagittal section through the mouth area. The section bisects the thickened esophagus and cuts the posterior part tangentially to show the musculature; 3, Sagital section of a highly contracted specimen showing the dorsal tubercles; 4, Outer area of a sagittal section showing the muscle layers and parenchymal fibers; 5, Lateral sagittal section of a contracted specimen showing excretory tubules; 6, Cross section of the dorsal area of a highly stretched specimen. Dorsal tubercles are flattened. Parenchymal cells and fibers are seen; 7, Medial sagittal section through the sperm duct and beginning of the seminal vesicle. The duct is somewhat bent and therefore seen in part tangentially; 8, Cross section just before the posterior end. There are no more dorsal tubercles. The excretory duct is simple and the two lateral nerves are visible; 9, Mature sperm cells from the seminal vesicle; 10, Cross section of the anterior body at the level of the nervous system. Gland cells around the esophagus; 11, Cross section around the middle of the ventral sucker which is withdrawn into the body. Retractor muscles are seen. There are two longitudinal nerves on each side with connecting commissures; 12, Cross section through the hindbody. The left intestinal cecum is making a bend; 13, From a sagittal section through the longitudinal axis of a highly rolled specimen. Part of the dorsal longitudinal musculature and its insertion in the surface, subcuticular cells and a flattened tubercle; 14, From a longitudinal section through the esophagus. In the upper part the longitudinal musculature is seen. Ducts and openings of the gland cells into the esophagus; 15, Excretory ducts with flame cells and nuclei from a section through the anterior body; 16, Frontal section through the duct connecting the testes with the gynecophoral canal beneath; 17, Part of a testis; 18, Reconstruction of the nervous system. The esophagus with its glands. Four testes with their connecting duct.

Figures 19 to 30, Females. 19, Cross section through the body at the level of the nervous system. Esophagus with its glands strongly pulled ventrally; 20, The same, somewhat further back through the esophageal glands; 21, Cross section at the level of the ventral sucker. The two intestinal ceca are quite separated; 22, Cross section just through the genital pore; 23, Cross section between the genital pore and the ootype. Intestinal ceca and uterus are seen; 24, Cross section through the common vitelline duct that contains an oocyte. Intestinal ceca very narrowed. Vitelline cells visible; 25, Section through the sphincter of the ovary. Intestinal ceca considerably larger. Oviduct and vitelline duct; 26, Frontal section through the end of the ovary and ovarial sphincter. Oviduct with sperm cells. Vitelline duct.*, transition between the sphincter and oviduct; 27, Sagittal section through the ootype and shellgland (= Mehlis' gland). Many sperm cells are with the oocytes and vitelline cells; 28, Cross section through the hindbody with unpaired intestinal cecum and vitelline glands. Vitelline duct ventrally; 29, Developmental stages of the oocytes; 30, mature oocyte in the oviduct.

Abbreviations: BR, ventral groove; BSN, ventral sucker; C, tegument; CC, cerebral commissure; DE, ejaculatory duct; DG, vitelline duct; DM, diagonal musculature; Dr, gland cells; DSt, vitelline gland; Ex, excretory vesicle; GC, cerebral ganglion; GZ, ganglion cells; H, anterior testes; HLN, major longitudinal nerve of hindbody; J, intestinal ceca; LM, longitudinal musculature; MSD, opening of shell gland; NDA, anterior dorsal nerve; NDP, posterior dorsal nerve; NLA, anterior lateral nerve; NS, nervous system (brain); NVA, anterior ventral nerve; NVP, posterior ventral nerve; Od, oviduct; Oe, esophagus, PG, genital pore; PM, parenchymal muscle; PZ, parenchymal cells; RM, circular muscle; RN, circular nerves; RSN, retractor muscles of the acetabulum; SB, seminal vesicle; SD, shell gland; Sp, spermatozoa; Sph, sphincter of ovary; SZ, subtegumental cells; Ut, uterus; VOD, common yolk-viteline duct; Z, dorsal tubercles of the male. *Source:* Looss 1895.

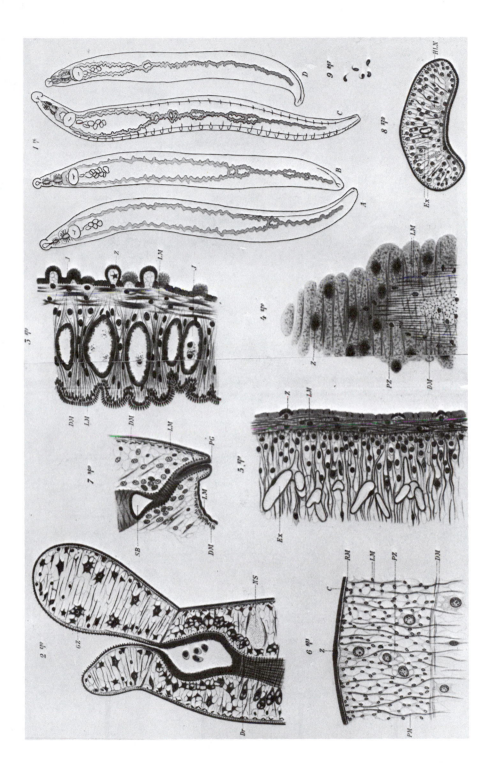

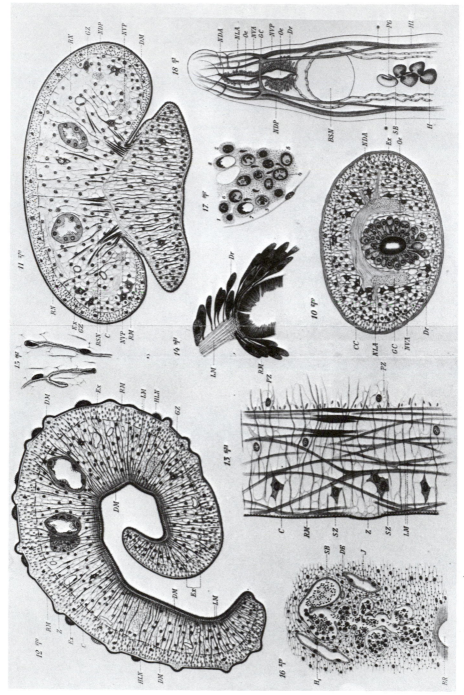

Figure 3–1 (continued)

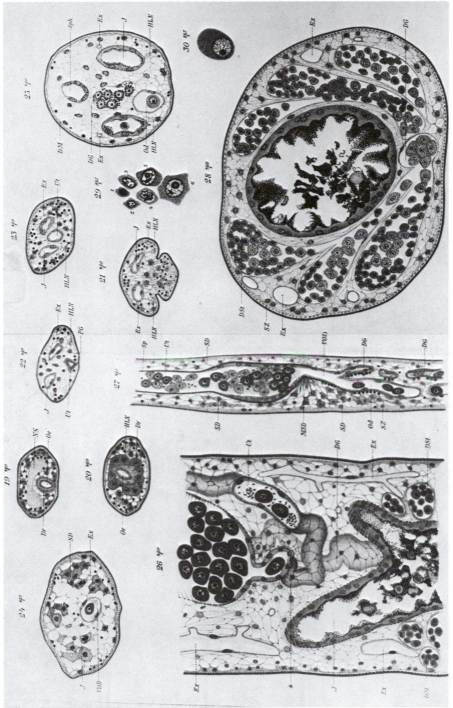

Figure 3–1 (continued)

TABLE 3–2 Representative Dimensions of Mature Adult Male Schistosomes

Species/ Origin	Host	Length		Width		Number	Reference Note
		Range	Mean	Range	Mean		
S. mansoni							
Puerto Rico	Rabbit, Monkey	6.4–9.9		0.36–0.58			1
Brazil-BH	Mouse	4.6–8.3	6.9	0.14–0.36	0.27	22	2
Brazil-SJC	Mouse	3.9–8.6	5.6	0.16–0.30	0.21	23	3
Brazil-Ba	Human		9.5[a] / 7.5[b]				
S. rodhaini							
Uganda	Mouse	2.0–4.4	3.1		0.25		4
	Mouse	3.1–5.1	4.0		0.25		
Zaire	Mouse		6.5		0.4		5
S. haematobium							
Sudan	Hamster	–20.0	13.5				6
S. Africa	Hamster	–17.0	11.0				
Iran	Mouse	4.8–10.9	8.22			56	7
	Mouse	8.3–15.2	10.83			10	
S. bovis							
Corsica	Cow	9–20					8
	Guinea Pig	8–10					
	Mouse	4.5–6.5					
S. nasalis							
India	Buffalo	6.7–11.6		0.23–0.56		30	9
S. lieperi							
Zambia	various	5.5–7.5					10
S. incognitum							
India	Mouse	2.4–4.7	4.06				11
	Dog	6.4–8.3	7.7				
	Pig	5.1–7.6	6.7				

Species / Locality	Host						
S. intercalatum							
Cameroon	Hamster	11.5–14.5	13.0	0.31–0.5			12
Lower Guinea	Hamster	9.0–16.0	11.2				
	Gerbil	8.3–14.3					
	Rat	4.2–5.0	4.6				
	Mouse	11.8–12.8	12.2				
S. japonicum							
Philippines	—	6–9.7	7.7			50	13
Japan	Human	8–21	8.38				14
	Cattle		16.				
	Cat	7–12	10.43				15
S. mekongi							
Laos	Dog	9.9–18.0	15.2	0.38–0.63	0.58	5	16
China	Mouse	11–14.5	13.				17
S. sinensium							
Thailand	Mouse	2.3–5.3	3.67	0.12–0.29	0.21		18
S. malayensis							
Malaysia	Mouse	4.3–9.2	6.82	0.24–0.43	0.32		19
Schistosomatium douthitti							
Michigan	Rat	1.9–4.53	3.07			70	20
	Peromyscus	2.9–5.27				9	
	Microtus	4.1–6.3					
	Mouse	2.5–5.2					
Schistosomatium pathlocopticum							
Massachusetts	Mouse	5.6–11.8		0.4–0.9			21
Heterobilharzia americanum							
Texas	Raccoon		9.06				22
	Raccoon		17.57				
	Mouse		4.				
	Dog		22.03				

TABLE 3–2 Representative Dimensions of Mature Adult Male Schistosomes (*continued*)

Species/ Origin	Host	Length		Width			Reference Note
		Range	Mean	Range	Mean	Number	
Dendritobilharzia pulverulenta Iowa	Ducks	2.9–13.3	6.65	0.4–1.79	1.01	110	23
Ornithobilharzia sp. Israel	*Larus*	3.3–16.5	9.7	0.6–0.8	0.7		24
Microbilharzia variglandis Rhode Island	Birds	2.9–4.2			0.5		25
Austrobilharzia terrigalensis Australia	Birds	2.4–4.2		0.26–0.38		5	26

[a]Asymptomatic

[b]Symmers Fibrosis

Abbreviations: Ba, Bahia; BH, Belo Horizonte; SJC, Sao Jose dos Campos

Reference notes: 1, Faust et al. 1934; 2, Magalhaes and de Carvalho 1973; 3, Cheever 1968; 4, Fripp 1967; 5, Brumpt 1931; 6, James and Webbe 1975; 7, Sabha and Malek 1977b; 8, Brumpt 1930a; 9, Dutt 1967; 10, Le Roux 1955; 11, Sinha and Srivastava 1956; 12, Wright et al. 1972; 13, Garcia 1976; 14, Fujinami 1907; 15, Katsurada 1904; 16, Iijima et al. 1971; 17, Moloney and Webbe 1983; 18, Greer et al. 1989; 19, Greer et al. 1988; 20, Price 1931; 21, Taylor 1970; 22, Goddard and Jordan 1980; 23, Ulmer and vandeVusse 1970; 24, Witenberg and Lengy 1967; 25, Stunkard and Hinchcliffe 1952; 26, Rohde 1977.

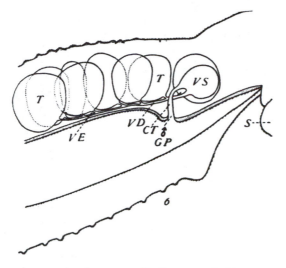

FIGURE 3–2 The male reproductive system in *S. mansoni*. CT, cirrus tube; GP, genital pore; S, ventral sucker (acetabulum); T, testis; VD, vas deferens; VE, vas efferens; VS, seminal vesicle. *Source:* Faust, Jones and Hoffman 1934.

the work of Leiper (1916, 1918) it is likely that *S. haematobium* and *S. mansoni* were confused. Looss (1895) cites the seminal work of Bilharz, who found 4–5 testes, and Leuckart, who described 6–8, characteristic of the two species, respectively.

The male system in *Schistosoma* consists of a series of spherical to oblate testes, commonly 4 to 8 in number (Table 3–3), located between the intestinal ceca in the sagittal plane posterior to the acetabulum and dorsal to the anterior part of the gynecophoral canal. The description of Severinghaus (1928) remains unsurpassed for clarity and succinctness. The testes may be linear or zigzag in arrangement. Males in some other genera (e.g., *Dendritobilharzia, Ornithobilharzia*) may possess hundreds of testes.

In *Schistosoma* the median ventral area of each testis forms an irregularly defined cavity or chamber to collect matured sperm; a short vas efferens extends ventrad, connecting seriatim with a vas deferens that leads to a seminal vesicle located in front of the anteriormost testis. From the ventral margin of the seminal vesicle a duct variously termed the ejaculatory or cirrus duct conducts sperm to the male genital opening, located in the median ventral plane at the anterior end of the gynecophoral canal just posterior to the acetabulum. The vas deferens and ejaculatory duct are devoid of surrounding glands (Otubanjo 1980), and it is difficult to understand the significance of the term "prostate" used by some authors. Although Severinghaus (1928) and some other earlier authors believed otherwise, most investigators report that no extensible or protrusible penis-like cirrus or cirrus sac is present. For example, a muscular cirrus tube was said to be absent in *S. rodhaini* (Faust and Meleney 1924), *S. bovis* (Lengy 1962b) and *S. haematobium* (Lindner 1914). Faust and Meleney (1924) described a median

TABLE 3–3 Testes Number in Schistosomes

Species/ Origin	Number		Observations	Reference Note
	Range	Mean		
S. mansoni				
Puerto Rico	6–9	usually 8		1
	6–11	7.91	n = 100	2
Liberia	0–19	6.87	Mouse infection. n = 53	3
E. Africa	2–14	6.62–9.86	12 separate isolates	4
Brazil-BH	5–12	8.3	n = 83	5
Brazil-SJC	5–12	7.1	n = 76	
Egypt	6–12	7.86	n = 100	6
Tanzania	4–8	5.85	n = 100	
S. rodhaini				
Uganda	4–8			7
Kenya	6–11	usually 7–9		8
S. haematobium				
Egypt	0–8		74%–4; 18%–5; 0.2%–7; n = 1,704	9
Tanzania	1–9		44%–4; 26%–5; 15%–6; 5%–7; n = 2,015	
S. bovis				
Israel	3–5	usually 4		10
S. intercalatum				
Cameroon	3–6	4.22	n = 82	11
	2–7		usually 4	12
Zaire	3–6	4.02	n = 92	
S. incognitum				
India	2–7		67% have 6 or 7	13
S. nasalis				
India	3–4			14
S. japonicum				
4 areas	1–14		90.5% had 7; n = 4,064	15
3 areas			usually 7	16
China	7		always 7	17
S. mekongi				
Laos	3–8	6.5		18
S. sinensium				
Thailand	4–8		usually 5–6	19
S. malayensis				
Malaysia	6–8			20
Schistosomatium douthitti				
Michigan	15–36			21
Schistosomatium pathlocopticum				
Massachusetts	14–18	usually 16		22
Orientobilharzia harinasutai				
Thailand	55–7–		Single row	23
Austrobilharzia variglandis				
Hawaii	variable		2 wk old–13; 4 wk–20; older–23	24

TABLE 3–3 *(continued)*

Species/ Origin	Number		Observations	Reference Note
	Range	Mean		
Austrobilharzia terrigalensis				
Australia	17 to 24		Approximately	25
Dendritobilharzia pulverulenta				
Iowa	100–170			26
Trichobilharzia ocellata				
Canada	57–76	67		27
Trichobilharzia cameroni				
Canada	80–100			28
Trichobilharzia physellae				
Michigan	96–110			29
Ornithobilharzia canaliculata				
Israel	60–115	86		30
Griphobilharzia amaena				
Australia	1		At posterior end of body	31

Testes number by Genus		
Genus	Yamaguti (1971)	Farley (1971)
Austrobilharzia	18–20	11–60
Bilharziella	ca. 110	ca. 110
Bivitellobilharzia	over 40	45–85
Chinhuta	70–80	Syn of *Bilharziella*
Dendritobilharzia	Numerous	Numerous
Gigantobilharzia	Numerous	To well over 300
Griphobilharzia	1[a]	1[a]
Heterobilharzia	70–83	To 83
Macrobilharzia	to 250	Over 200
Microbilharzia	ca. 20	Syn of *Austrobilharzia*
Orientobilharzia	37–80	37–80
Proschistosoma	3–5	Not discussed
Pseudobilharziella	Numerous	Syn of *Trichobilharzia*
Schistosoma	Less than 10	Not discussed
Schistosomatium	14–18	15 to 36
Sinobilharzia	ca. 65	?
Trichobilharzia	Numerous (to 160)	—

[a]Not mentioned by Yamaguti or Farley

Reference notes: 1, Faust et al. 1934; 2, Saoud 1965; 3, Gönnert 1949; 4, Coles and Thurston 1970; 5, Magalhaes and de Carvalho 1973; 6, Saoud 1966c; 7, Fripp 1967; 8, Saoud 1966b; 9, James and Webbe 1973; 10, Lengy 1962; 11, Bjorneboe and Frandsen 1979; 12, Wright et al. 1972; 13, Sinha and Srivastava 1956; 14, Dutt 1967; 15, Hsü and Hsü 1957; 16, Mao and Li 1948; 17, Faust and Meleney 1924; 18, Iijima et al. 1971; 19, Greer et al. 1989; 20, Greer et al. 1988; 21, Price 1931; 22, Tanabe 1923; 23, Kruatrachue et al. 1965; 24, Chu and Cuttress 1954; 25, Rohde 1977; 26, Ulmer and vandeVusse 1970; 27, Bourns et al. 1973; 28, Wu 1953; 29, McMullen and Beaver 1945; 30, Witenberg and Lengy 1967; 31, Platt et al. 1991.

ventral outlet for the male system in *S. japonicum,* which enlarges at copulation and is applied to the female genital orifice to transfer sperm. In *S. bovis,* Kuntz et al. (1979) described and figured a "tripartite" gonopore, which has not been discussed elsewhere or for other species.

In males of genera other than *Schistosoma,* a genital papilla may be present. For *Trichobilharzia ocellata,* Bourns et al. (1973) described a "stout" papilla protruding continuously from the anterior margin of the gynecophoral canal, generally from the left wall.

Development. The ontogeny of the male reproductive system in schistosomes has not been studied in detail, although Cort (1921) described "traces" in *S. japonicum* schistosomula 1 to 2 mm long.

Testes

Structure. In some species or local isolates there may be a tendency to lobulation, as in *S. japonicum* (Faust and Meleney 1924) or in *S. bovis* (Lengy 1962b). In *S. mekongi* Voge et al. (1978a) remarked that testes may be smooth or deeply lobed in different worms, even within the same individual host. In *S. rodhaini* the anteriormost testis is usually largest (Fripp 1967); in *S. bovis,* the posteriormost (and sometimes anteriormost) testis is larger (Lengy 1962b). Each testis is bound by a dense heterogeneous basal lamina within a layer of circular muscle fibers (Otubanjo 1981b). Germinal cells were described by early authors such as Lindner (1914) and Severinghaus (1928), but have not been the subject of recent scrutiny. Within the testis the developmental stages are jumbled together so that it is difficult to distinguish cell types (Gönnert 1947). Two types of nongerminal cells possibly possessing nutritive, supportive, or phagocytic functions have been described on an ultrastructural level by Otubanjo (1981b), who suggested that they may help to regulate the production of spermatozoa.

Number. The distribution of testis number in males of various species and isolates is presented in Table 3–3. The number is usually fairly constant for each species; for example, in *S. japonicum* the modal number for virtually all isolates is seven. In the case of *S. mansoni* the mean number ranges from about 6.6 to more than nine in various population cohorts (reviewed briefly by Saoud 1966a); in the case of worms from Mwanza, Tanzania, Saoud (1965) found that 60% of males had fewer than 7 testes, bringing the mean number in that population close to that of the *S. haematobium* complex. Following up on this observation, Coles and Thurston (1970) investigated testis number in 12 East African isolates of *S. mansoni,* concluding that highly significant differences ($p = .001$) exist among isolates between, and even within, individual villages. There was as much variation within a given area as among isolates from different geographic regions, demonstrating that testis number is not a valid criterion to distinguish among nominal "strains." The possible role of the host in influencing testis number was studied by

Saoud (1966a), who concluded that testis number in males from mice and hamsters was significantly different within certain isolates of *S. mansoni,* but not in others. Zanotti et al. (1982) found that the mean number of testes in males was somewhat higher (8.13 vs. 7.80) in mice with bisexual infections than in those with unisexual male infections of Brazilian *S. mansoni.*

Although some mammalian schistosomes (e.g., *Orientobilharzia*) have many dozens of testes, the schistosomes of birds have, in general, far more testes, commonly over 100 (Wilson et al. 1978, Yamaguti 1971, Skriabin 1951). Moreover, the testis number may increase with the age of the worm, at least in *Austrobilharzia variglandis* in Hawaii (Chu and Cutress 1954).

Anomalies. Anomalies in the testes have been described, particularly in *S. japonicum* where Hsü and Hsü (1957), comparing worms from China, Japan, Taiwan, and the Philippines described variations in arrangement, number, and location. A strong genetic component was suggested, as 9.8% of Taiwanese males in rabbits possessed unusual features, whereas only 1.5% of males from Japan did so in the same host. Ho (1962) described supernumerary testes as well as testes in unusual locations: 15 of 6,000 worms from China had extra sets. Supernumerary testes were found also by Moore (1955) in *S. japonicum* of Taiwanese origin. In *S. mansoni* from Egypt, Najim (1951) described a worm that contained 9 testes in the normal position, plus 5 more in an isolated, more posterior location. Saoud (1965) found that 10% of males of the "Wellcome strain" originating in Egypt had supernumerary testes, while none of 100 males of Puerto Rican or Tanzanian origin had any. The presence of oocytes in the testes of *S. haematobium* was described by Deters and Nollen (1976).

The anticancer drug procarbazine was found by Basch and Clemens (1989) to cause reversible and almost complete destruction of testes in *S. mansoni.* On average the number of regenerated testes was one fewer than the number present in untreated worms.

Spermatogenesis

Spermatogenesis in trematodes has been reviewed in detail by Mohandas (1983), who cited almost 100 references on the process in Aspidogastrea, Monogenea, and Digenea. The cytology of gametogenesis in schistosomes has been studied since the early years of this century, in large part to help understand the biology of sexual differentiation in this group. Authors generally agree on details of gamete production and structure, which show few unique features.

The pioneering work of Lindner (1914) referred nominally to *S. haematobium,* but considering the state of taxonomy in 1914, particularly among the German school of parasitologists, it appears possible that *S. mansoni,* or perhaps even another species, could have been used. Preserved material of variable quality from humans from Egypt was studied, as well as some *S.*

bovis from Sudanese cattle. Proceeding under the tutelage of Richard Hertwig and Richard Goldschmidt, Lindner found that spermatogonia, primary and secondary spermatocytes, and spermatids, all of which he described in detail, were broadly distributed within the stroma of the testes.

Spermatogenesis in *S. japonicum* was studied by Faust and Meleney (1924), and in greater detail by Severinghaus (1928) a few years later. The latter embarked on "a tedious cytological study" extending over 2 years, using several thousand appropriately fixed males. Germ cells were found to originate in sheets from the wall of the testis, developing near the periphery into spermatogonia. Subsequent stages were difficult to interpret, as daughter cells separated immediately after division to migrate through the testis (also mentioned by Lindner 1914): there was "not the slightest indication of a regional distribution of developmental stages." The meiotic divisions were described in great detail, leading to spermatids and eventually to mature sperm cells in a conventional manner. Male worms fixed 26 days after infection showed only spermatids, as mature sperm cells were not yet present.

In a more modern study, Den Hollander and Erasmus (1984) followed the incorporation of tritiated thymidine into *S. mansoni,* finding a highly generalized distribution of radioactivity in the total body tissues to about 35 days after infection, then declining to about 85 days after infection. In both paired and unisexual males older than 35 days, ^{3}H-hymidine incorporation was observed principally in the testes, indicating a continuing high degree of DNA synthesis in reproductive tissues with only isolated activity in somatic sites. At 45 days after infection, autoradiography indicated that the densely labeled cells were always in a peripheral position in the testes, with the central area generally unlabeled.

For *S. mansoni,* Niyamasena (1940), in Vogel's laboratory, found some difficulty in distinguishing stem cells from spermatogonia, located at the edges and in the interior of the testes. In fully mature males, only resting (not dividing) nuclei were observed, and division stages were detectable, although still relatively rarely, only in younger males. Meiotic and sperm maturational stages were found essentially as described by Lindner (1914), Faust and Meleney (1924), and Severinghaus (1928). Kitajima et al. (1976) reported that in late spermatocytes of *S. mansoni* the Golgi complex is weak or absent, perhaps associated with the absence of an acrosome in the mature sperm. The sustentacular cells of the testes, described by Otubanjo (1981a), presumably regulate sperm production and are closely associated with spermatids and maturing sperm.

For *Schistosomatium douthitti* Nez and Short (1957) traced spermatogenesis, finding that each primary spermatogonium gave rise by 3 mitotic divisions to 8 primary spermatocytes. Two maturation divisions subsequently produced a coherent cluster of 32 spermatids, in contrast to *Schistosoma,* where the various daughter cells separated chaotically. This pattern is more reminiscent of spermatogenesis in hermaphroditic trematodes. Moreover, a zonation of cells in the testicular follicles of *S. douthitti* is unlike the condition in *Schistosoma,* further indicating an early phylogenetic split between these two genera.

Structure of the Mature Sperm

Sperm are produced by both paired and unpaired males in mixed infections, and by unisexual males, at least in the major human schistosomes. Worms cultured in vitro also produce sperm cells, as recorded by Clegg (1959) and Irie et al. (1983) for *S. mansoni,* Smith et al. (1976) for *S. haematobium,* and Yasuraoka et al. (1978) for *S. japonicum.*

The fine structure of *S. mansoni* sperm was best described by Kitajima et al. (1976). In general morphology these cells are typical of animal sperms but atypical of most platyhelminthes, whose sperm are commonly biflagel-late (as in *Schistosomatium douthitti*) or pointed at both ends. There is a head measuring 8 by 2 μm, lacking an acrosome, rounded anteriorly and tapering posteriorly. A prominent anterior mass of undifferentiated mito-chondria lies beneath the plasma membrane, in a cuplike depression in the nucleus. The nucleus itself is largely electron-opaque, but contains electron-lucent patches of Feulgen-negative material. A single layer of microtubules is arrayed longitudinally between the periphery of the large nucleus and the plasma membrane of the sperm head. The single flagellum is about 20 μm in length, described by Kitajima et al. (1976) as of the 9 + 0 configuration, containing 9 doublets and lacking the central pair of tubules. In *S. mansoni* adult males completely cultured in vitro by the method of Basch (1981a), sperm were essentially identical in ultrastructure (Irie et al. 1983) to those from worms taken from animal hosts.

Justine and Mattei (1981) have argued that schistosome sperm flagella are not truly of the 9 + 0 type because they possess a nonmicrotubular, electron-dense central element that contains glycogen. They have desig-nated these sperm flagella as a new 9 + "1" to emphasize the unique central element.

In *Schistosomatium douthitti* the sperm are quite different from those of the genus *Schistosoma* (Nez and Short 1957), resembling those of many her-maphroditic trematodes and some turbellaria (Hyman 1951). They consist of a thin head about 30- to 35-μm long, sometimes appearing ribbonlike, and two short flagella less than half the body length.

Gynecophoral Canal

The surface configuration of this structure has been revealed by means of scanning electron microscopy. SEM studies have been made on males of the following species: *S. bovis* (Kuntz et al. 1979); *S. haematobium* (Hicks and Newman 1977, Kuntz et al. 1976, Leitch et al. 1984); *S. japonicum* (Ma and He 1981, Sakamoto and Ishii 1977, Voge et al. 1978); *S. malayanum* (Sobhon et al. 1983); *S. leiperi* (Southgate et al. 1981); *S. mansoni* (Basch and Basch 1982, Evers et al. 1983, Price et al. 1978, Race et al. 1971, Senft et al. 1978); *S. margrebowiei* (Ogbe 1982); *S. mattheei* (Tulloch et al. 1977); *S. mekongi*

(Voge et al. 1978c); *S. sinensium* (Kruatrachue et al. 1983b); and *S. spindale* (Kruatrachue et al. 1983a). Although these descriptions vary to some extent, the surface of the gynecophoral canal is usually densely strewn with irregularly directed spines less robust than those covering the acetabulum, which presumably function to help maintain the female within the canal (Fig. 3–3). The underlying tegument may be thrown into small folds and ridges, and pitted or honeycombed with uniformly distributed pores (Price et al. 1978, Voge et al. 1978b). The spines are interspersed with sensory bulbs bearing elongated papillae; these sensory receptors are typically more abundant at the anterior and posterior ends and near the lateral margins than in the center of the gynecophoral canal. The edges of the canal also commonly possess many small, blunter spines.

Functional studies are far fewer than morphological descriptions of the gynecophoral surface. Senft (1968) believed that nutrients, hormones, or pheromones could be transferred from male to female schistosomes across this membrane, and suggested transfer experiments involving radioactively labeled substrates; some such experiments have since been done, as described in Chapter 4. Senft et al. (1978) have suggested that in *S. mansoni* this area is a site of active lipid secretion, showing numerous globules that appear to be lipid droplets.

Aronstein and Strand (1985) have described a monoclonal antibody immunoreactive against a glycoprotein antigen differentially expressed on the gynecophoral canal of mature male *S. mansoni*. Such specific monoclonal antibodies promise to be a useful tool in exploring schistosome biology.

The Female Schistosome

Female schistosomes, particularly those parasitizing birds, exhibit considerable variation in shape, from elongated and filiform, as in *Trichobilharzia,* to flattened and leaflike, as in *Dendritobilharzia* (Fig. 1–11). The shape of the female was a prime consideration of Price (1929), in the erection of two subfamilies: the Schistosominae with slender, cylindrical females, and the Bilharziellinae with females similar to males. Mammalian species in general are less variable but still exhibit a range of forms, from the familiar cylindrical females of *Schistosoma* to the broader, flattened females in *Schistosomatium.*

Dimensions of some adult female schistosomes are listed in Table 3–4; stated sizes should be considered only approximate, as dimensions can vary because of biological and technical factors discussed earlier.

The external surface of those female schistosomes that have been observed by scanning electron microscopy is simpler and more uniform than that of the male (Basch and Basch 1982, Kuatrachue et al. 1979, Leitch et al. 1984, McLaren 1980, Price et al. 1978, Race et al. 1971, Senft et al. 1978, Silk et al. 1970, Sobhon et al. 1983, Voge et al. 1978b, c). This consists for the most part of a relatively featureless tegument with delicate transverse ridges bearing both ciliated and unciliated, presumably sensory, papillae.

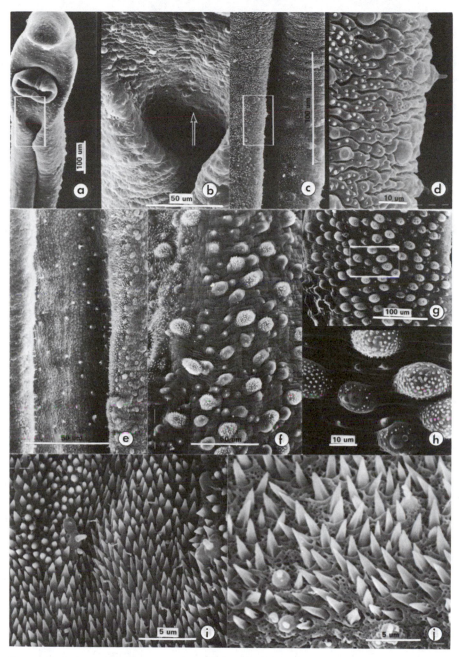

FIGURE 3–3 Scanning electron micrographs of adult male *S. mansoni* cultured in vitro, 3 or more months old. a, ventral view of anterior body, with beginning of gynecophoral canal; b, detail of a genital aperture indicated by arrow; c, edge of gynecophoral canal at midbody; d, detail showing blunted spines and sensory papillae; e, view straight down into gynecophoral canal in midbody; f, lateral midbody with spinous tubercles. Ventral edge of gynecophoral canal to the left; g, lateral midbody, dorsal to the left; h, detail of g, tubercles with spines and sensory papillae; i. Spinous area near edge of gynecophoral canal, with sharp and blunt spines and sensory papillae; j, another specimen, edge of gynecophoral canal, midbody. Note pitted tegument.

TABLE 3–4 Representative Dimensions of Mature Adult Female Schistosomes

Species/ Origin	Host	Length		Width		Number	Reference Note
		Range	Mean	Range	Mean		
S. mansoni							
Puerto Rico	Rabbit, Monkey	7.2–14		0.2–0.38			1
Brazil-BH	Mouse	5.1–10.9	7.6	0.03–0.06	0.04	20	2
Brazil-SJC	Mouse	4.2–7.1	5.8	0.03–0.06	0.04	13	3
Brazil-Ba	Human		10.6[a] 8.2[b]				
S. rodhaini							
Uganda	Various	1.3–3.0	2.3		0.1	104	4
Zaire	Mouse	9–10.5			0.2		5
S. haematobium							
Iran	Mouse	7.5–19.4	12.94	0.235–0.338	0.268	10	6
S. bovis							
Corsica	Mouse 1	5.4–7.5	6.57			7	7
	Mouse 2	8–16.1	11.2			5	
Sudan	Cow	12–14	12.88			4	
Israel	Mouse	12.8–20.0	15.2	0.15–0.25	0.17		8
S. nasalis							
India	Buffalo	6.9–11.7		0.103–0.210			9
S. lieperi							
Zambia	Various	6.5–8.5					10
S. incognitum							
India	Mouse	3.0–3.14	3.05	0.072–0.13	0.132		11
	Dog	4.8–7.6	6.5	0.1–0.112	0.11		
	Pig	4.3–7.1	5.93	0.1–0.112	0.11		

Species / Origin	Host						Ref.
S. intercalatum							
Cameroon	Hamster	13–24		0.2–0.25			12
S. mattheei							
S. Africa	Sheep		21.8		0.25		13
S. japonicum							
Philippines	—	9–14				50	14
	Human	12–17					
	Cattle	12–23					
	Cat	8–12					
Japan	—	9–14.5	11.1				15
Japan	Mouse	8–9.8					17
China		varies with age	11.6	0.15–0.3	0.23	5	18
China							19
S. mekongi							
Laos	Dog	16.3–20.1	17.8	0.28–0.3	0.28		20
S. sinensium							
Thailand	Mouse	3.0–5.7	4.32	0.08–0.15	0.12		21
S. malayensis							
Malaysia	Mouse	6.5–11.3	9.16	0.15–0.28	0.21		22
Schistosomatium douthitti							
Michigan	White Rat	1.6–3.6	2.48			50	23
	Peromyscus	2.5–4.3				11	
	Microtus	1.4–4.8				18	
	Mouse	1.1–3.7				9	
Schistosomatium pathlocopticum							
Massachusetts	Mouse	4.5–10.2		0.18–0.38			24
Orientobilharzia harinasutai							
Thailand	Rodents	1.7–3.2		0.06–0.1			25

TABLE 3–4 Representative Dimensions of Mature Adult Female Schistosomes (*continued*)

Species/ Origin	Host	Length		Width		Number	Reference Note
		Range	Mean	Range	Mean		
Dendritobilharzia pulverulenta Iowa	Various		6.73		1.00		26
Trichobilharzia cameroni Canada	—	3.8–4.9		0.052–0.059			27
Ornithobilharzia canaliculata Israel	*Larus*	3.4–6.8	5.1	0.11–0.29	0.19		28
Microbilharzia variglandis Rhode Island	Birds	2.0–3.5		0.05–0.175			29
Austrobilharzia terrigalensis Australia	Birds	2.6–3.4		0.10–0.14			30

[a]Asymptomatic

[b]Symmer's fibrosis

Reference notes: 1, Faust et al. 1934; 2, Magalhaes and de Carvalho 1973; 3, Cheever 1968; 4, Fripp 1967; 5, Brumpt 1931; 6, Sahba and Malek 1977b; 7, Brumpt 1930a; 8, LeRoux 1949; 9, Dutt 1967; 10, LeRoux 1955; 11, Sinha and Strivastava 1956; 12, Wright et al. 1972; 13, van Rensburg and van Wyk 1981; 14, Garcia 1976; 15, Fujinami 1907; 16, Katsurada 1904; 17, Iijima et al. 1971; 18, Faust and Meleney 1924; 19, Moloney and Webbe 1983; 20, Iijima et al. 1971; 21, Greer et al. 1989; 22, Greer et al. 1988; 23, Price 1931; 24, Tanabe 1923; 25, Kruatrache et al. 1965; 26, Ulmer and vandeVusse 1970; 27, Wu 1953; 28, Witenberg and Lengy 1967; 29, Stunkard and Hinchcliffe 1952; 30, Rohde 1977.

Under higher magnification tegumental pores may be seen throughout. Near the posterior end there is commonly an area of long, sharp spines encircling the body except for the very posterior tip, which is marked by small excrescences of various kinds around the excretory pore.

Female Reproductive System

Hyman (1951) pointed out that in most animal species the ovary produces both the haploid ovum and associated nutritive materials. In the Platyhelminthes these functions are undertaken by separate organs: oogenesis occurs exclusively in the ovary; whereas vitelline cells, which appear primitively to be abortive homologs of ova, are given off in larger numbers by the vitelline glands (sometimes called vitellaria). A generalized diagram of the trematode female reproductive tract, from the monograph of Smyth and Halton (1983), is shown in Figure 3–4; specific patterns in numerous trematode

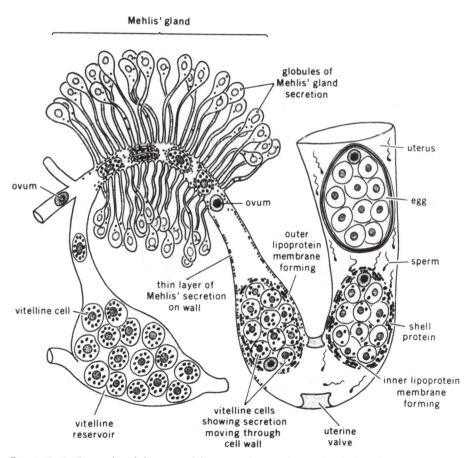

FIGURE 3–4 Generalized diagram of the central area of a trematode female reproductive system, based on *Fasciola*. *Source:* Smyth and Halton 1983.

families are described and figured by Ebrahimzadeh (1966), from whose review the schistosomes are strangely absent. The names applied to components, including cellular elements of the reproductive system, vary greatly among authors.

The schistosome female reproductive tract (Fig. 3–5) has been an object of study for more than a century. The system (presumably of *S. haematobium*) was described in general terms by Fritsch (1888) and in greater detail by Looss (1895); both referred to earlier observations on the subject by Bilharz. In Japan the system was discussed and beautifully illustrated by Fujinami (1907). The female reproductive system of *S. mansoni* is shown by Spence and Silk (1971a) and of *S. japonicum* by Vogel and Minning (1947). A clear verbal description of gross features of the system in *S. mansoni* is found in Faust et al. (1934).

In contrast to the paired ovaries present in most trematodes, schistosomes possess only one. The oviduct is modified proximally to form a dilated area, to which sperm cells have swum upstream from the distant female reproductive opening or gonopore. Vitelline glands, often elaborately divided, discharge through the vitelline duct into the oviduct, carrying oocytes newly released from the ovary, which is thereupon termed the ovovitelline (sometimes vitello-ovi-) duct. A Laurer's canal originates from this region in some members of the family, including *Heterobilharzia* (Lee 1962), *Dendritobilharzia* (Ulmer and vande Vusse 1970), *Trichobilharzia* (McMullen and Beaver 1945), *Schistosomatium* (El-Gindy 1951), and at least some species of *Ornithobilharzia* (Price 1929—old classification) but is absent in *Schistosoma*. The package consisting of oocyte, sperm, and vitelline cells is subsequently processed in a complex linear assembly line with the aid of glandular secretions and physical shaping, to become an egg.

Development. Ontogenetic development of the female reproductive system has been described in *S. japonicum* by Tada (1928), who identified a genital primordium in 8-day-old females, and by Cort (1921a), who found clearly formed anlage of the ovary, vitelline gland, Mehlis' gland, uterus, and genital pore in an 18-day-old worm 1.44 mm in length recovered from the liver of a mouse. Other authors who have studied development of the female system to a greater or lesser degree include Lengy (1962b) (for *S. bovis*), Faust et al. (1934), Clegg (1965b), Spence and Silk (1971a), Erasmus (1973) (for *S. mansoni*), and El-Gindy (1951) (for *Schistosomatium douthitti*). The latter described female genital primordia, consisting of 18 to 30 cells, in 2- to 4-day-old worms, and provided a detailed study of subsequent cell numbers and movements. According to Spence and Silk (1971a), who carried out an ultrastructural study of the entire tract, the female reproductive system is basically secretory. It is tegumentary in origin and syncytial in nature, with gradual changes in architecture from one region to the next. Erasmus (1973) has considered the female reproductive system of *S. mansoni* as three ultrastructurally distinct regions: the ovary and oviduct; the vitelline gland and its duct; and the uterus, ootype, and Mehlis' gland.

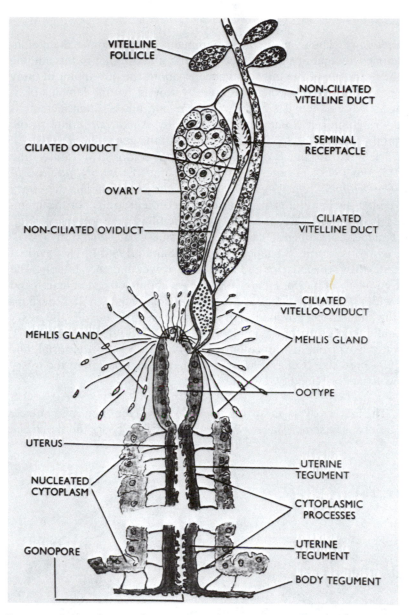

FIGURE 3–5 Female reproductive system in *S. mansoni,* diagrammatic. *Source:* Spence and Silk 1971b.

The Ovary

Gross Morphology. The ovary in most mammalian schistosomes is an elongate, sometimes conical or pear-shaped saccular gland, lying in the midline between the branches of the intestinal ceca at about the 40% point in body length (Fig. 3–6). In some species it is spirally wound: young females of *S. bovis* show as many as 9 twists or coils in the ovary, but in older specimens it tends to straighten out (Lengy 1962b). In *Schistosomatium douthitti* the ovary is spiral (Price 1931). Avian schistosomes of the genus *Trichobilharzia* possess ovaries that are linear or coiled in up to 4–6 loops (McMullen and Beaver 1945, Wu 1953); in other avian schistosomes (*Austrobilharzia, Bilharziella, Dendritobilharzia, Gigantobilharzia, Ornithobilharzia*) the ovary is usually coiled but may present a variety of configurations. In the mammalian parasite *Orientobilharzia harinasutai,* the blind anterior end is long, thin, and coiled, resembling a "tail" (Kruatrachue et al. 1965). In *S. rodhaini,* the ovary is spiral, S-shaped, or oval (Saoud 1966b). The oviduct emerges posteriorly; the anterior end is often narrower and ends blindly. The fully matured ovary of *S. japonicum* (Fig. 3–6) is about 400 µm in length and 150 µm in width (Cort 1921), that of *S. mansoni* somewhat smaller, perhaps reflecting the smaller egg production in that species. In *S. nasalis* the ovary is in the middle third of the body, 155 to 269 µm by 72 to 118 µm (Dutt 1967); in *S. bovis* it ranges from just preequatorial to definitely postequatorial, and is 560 by 125 µm in size (Lengy 1962b). The classic description of the ovary of *S. mansoni* was provided by Gönnert (1955a).

Histology. The mature ovary is surrounded by a thin layer of circular muscle outside of a fibrous basement layer, whose inner surface bears the thin basal

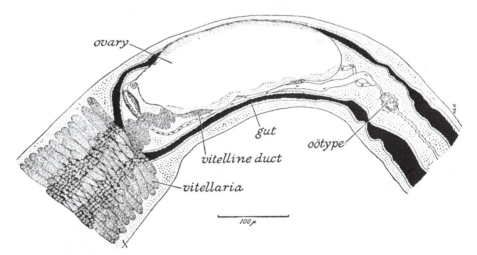

FIGURE 3–6 The central area of the female reproductive system in *S. japonicum. Source:* Bang and Hairston 1946.

epithelium of the ovary. Gönnert (1955a) described a covering epithelium, thinner at the blind anterior end and thicker at the base. The ovary is filled with cells in various stages of development, arrayed linearly from the smaller and more immature oogonia in the distant anterior area to oocytes ready for discharge near the oviducal junction. The closely-packed anterior blind portion is the major site of division (Erasmus 1973); posteriorly the cells are more loosely packed. The structure and organization of the ovary in mature *S. mansoni* was thoroughly described by Niyamasena (1940).

In unisexual female *S. mansoni* the ovary is small, often coiled, and relatively inconspicuous, measuring about 34 to 40 μm in width and 100 to 175 μm in length, and has been well described by Moore et al. (1954) and Erasmus (1973). Severinghaus (1928) found the ovary in unisexual *S. japonicum* to be barely present, with cells in a "heavy spireme" stage but no sign of typical egg cells.

Ultrastructure. Oogonia have a large irregularly shaped nucleus with clumped chromatin masses and a nucleolus, and a narrow rim of cytoplasm. Cytoplasmic organelles appear to vary with the stage of maturation and degree of metabolic activity (Spence and Silk 1971a). Immature ova, found by these authors to vary in size, are distinguished by production of cortical granules. As the cell matures the nucleolus increases in size and develops electron-lucent areas. The cell termed the "ovum" by Erasmus (1973), "mature oocyte" by various authors, and "primary oocyte" by Nollen et al. (1976), which is discharged from the ovary, lacks microvilli. There is usually a well-defined nucleolus and little nuclear chromatin; ribosomes are often associated with the nuclear membrane. The cytoplasm contains ribosomes, granular endoplasmic reticulum, mitochondria, and abundant Golgi material. Dense cortical granules, formed from the Golgi complexes, underlie the peripheral membrane (Spence and Silk 1971a, Erasmus 1973).

In *S. mansoni* cultured in vitro from cercariae to adults, Irie et al. (1983) have studied the development of cells in the ovary. Anterior oogonia appeared relatively normal, containing mitochondria, ribosomes, and granular endoplasmic reticulum. However, cells in the middle and posterior ovary become vacuolated and showed numerous degenerative changes. Although cortical granules were formed, egglike material in the ootype and uterus lacked germinal disc and shell and only abortive eggs were produced (Basch 1981; Irie et al. 1983).

Physiology and Function. Little is known about the control of cellular development and movement in the ovary of parasitic platyhelminthes (Nollen 1983) in general, and in schistosomes in particular. Nollen et al. (1976) incubated adults of *S. mansoni, S. japonicum,* and *S. haematobium* in tritiated thymidine for 2 hours and then implanted the worms surgically into hamsters. Nuclear label was taken up rapidly into oogonia at the anterior periphery of the ovary. After 3 days labeled cells appeared in the center anterior region of the ovary. Label was seen in the nuclei of primary oocytes in *S. japonicum* in 6 days, and in the other species in 7 days. Nollen (1981) found that extranuclear DNA was concentrated in association with mitochondria in oocytes of *S. mansoni* by electron microscope autoradiography. Den Hollander and Erasmus (1984) labeled *S. mansoni* females with tritiated thymidine, finding incorporation only in the anterior oogonial cells (termed by them oocytes); posterior regions of the ovary were always found devoid of label.

Drug effects on the ovary have been studied by several investigators. Antimonials were applied to *S. japonicum* by Bang and Hairston (1946) and Vogel and Minning (1947), who reported complete destruction of ovarian tissues. A similar result was found with miracil (Gönnert 1947). Niridazole caused a massive extrusion into the uterus of ovarian cells, which were then passed out by the worms (Striebel, 1969). Emetine, which was very destructive of the vitelline gland, had little effect on the ovary (Vogel and Minning 1947).

Anomalies. Supernumerary ovaries and an ovary with a spiral anterior portion were described in *S. japonicum* by Ho (1962). In *S. mansoni,* Moore et al. (1954) also found specimens with masses of ovarian tissue separate from the main ovary, and a specimen with an unusually long ovary.

The Vitelline Gland

Gross Morphology. The vitelline gland, sometimes called vitellaria, generally extends from the ovary to the posterior extremity of the typical female schistosome, occupying about two-fifths to two-thirds of the body. Numerous vitelline follicles connect through minor ducts to a vitelline duct, which eventually opens into the oviduct. In some species such as *S. douthitti* the vitelline follicles ventral to each of the paired intestinal ceca empty into a separate lateral duct, which unite into a common vitelline duct before connecting with the oviduct. The vitelline duct of *S. mansoni* was closely observed by Gönnert (1955a). Starting near the posterior end of the female worm, the duct proceeds anteriad on the ventral side until the gut forks just behind the ovary, then loops forward until shortly before the ootype. At this point the relatively broad lumen of the vitelline duct narrows so that the mature vitelline cells must lie in a row like a string of pearls. A very thin passage with thick walls intervenes, and the opening to the oviduct is formed into a valve, presumably to prevent reverse flow of vitelline cells.

The presence of scattered vitelline material in male *S. mansoni* has been observed by various authors, as mentioned in the section on hermaphroditism in this monograph. A histochemical and ultrastructural study of this phenomenon has found these cells in males to be essentially identical with mature vitelline cells of normal *S. mansoni* females (Shaw and Erasmus 1982). The extent to which vitelline material is found in male schistosomes of other species is unknown.

Histology. Vitelline follicles have a zonal structure roughly demarking groups of cells of the same stage of development. According to Gönnert (1955a), vitelline cells are haploid, further suggesting a common origin of the vitelline gland and ovary. Following treatment with Miracil D, some oocytes were seen by Gönnert (1955a) in the regenerating vitelline gland, but it was not certain whether these had originated in the vitelline gland or migrated in from the ovary. Immature cells cluster around the periphery of the follicle while the more mature cells migrate to the area proximal to the vitelline duct, into which the differentiated cells are extruded (Doughty 1975).

Histochemistry. Fast red B salt, a diazo dye that reacts with phenolic substances, has become the standard for demonstration of phenolic material since its suggestion by Johri and Smyth (1956). Reactivity to certain phenolic-specific dyes was found by Piva and deCarneri (1961) to begin between the 35th and 40th day of infection in paired *S. mansoni* of Liberian origin, and to remain stable for at least 2 months thereafter. Phenols and phenolase were found by a battery of methods only within globules of mature vitelline cells in *S. japonicum* and in the shells of eggs in utero; neither schistosomula nor male worms showed any histochemical reaction (Ho and Yang 1973). Phenol oxidase activity was localized to discrete packets within vitelline globules by Seed et al. (1978) on the basis of fluorescent histochemical studies. An abundance of alkaline phosphatase, but no acid phosphatase, was found in vitelline cells of *S. mansoni* by Piva and Grimley (1966).

The autofluorescence of vitelline material and egg shells is readily demonstrable with an ultraviolet light microscope, eliminating the necessity for histochemical staining.

Ultrastructure. A general description of vitelline cells was provided by Spence and Silk (1971a), but the most detailed studies are those on *S. mansoni* by Erasmus and his group in Wales. The development of vitelline cells was divided into four maturational stages by Erasmus (1975b) in a definition refined by Erasmus and Popiel (1980). Undifferentiated Stage 1 cells, at the lobular periphery, possess a large nucleus and nucleolus, and clusters of ribosomes are attached to the outer nuclear membrane (Fig. 3–7). The electron-lucent cytoplasm contains ribosomal clusters, Golgi complexes, and mitochondria, but little glycogen. Stage 2 cells show the beginning of synthetic activity. They have a greater cytoplasmic volume: ribosomal clusters and granular endoplasmic reticulum are abundant. Stage 3 cells demonstrate active protein synthesis. They have no ribosomal complexes but much more GER; abundant Golgi complexes are associated with early stages of vitelline droplet formation. Mature stage 4 cells contain patches of nuclear heterochromatin; the cytoplasm is dense, with packed ribosomes, GER, droplet-producing Golgi complexes, lipid droplets, and vitelline droplets. These were up to 2 μm in diameter, made up of smaller globules, and arrayed near the cell periphery. Erasmus and Davies (1979) showed that mature vitelline cells also contain calcareous corpuscles, probably containing a phosphorus

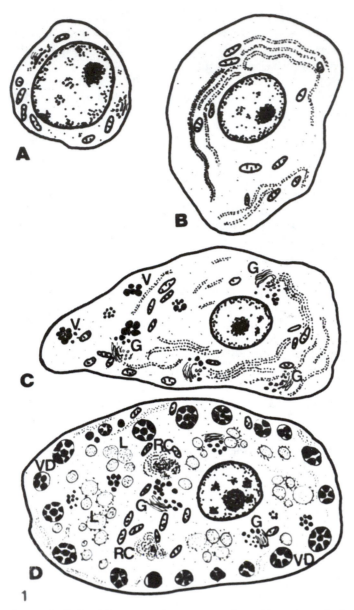

FIGURE 3–7 Semidiagrammatic representation of four stages in the development of the vitelline cell of *S. mansoni*. A, Stage 1. B, Stage 2. Note appearance of the endoplasmic reticulum. C, Stage 3. The Golgi complexes (G) have begun to secrete the vitelline globules (V); D, Stage 4. This is the "mature" vitelline cell with vitelline droplets (VD), ribosomal complexes (RC), Golgi complexes (G), and abundant lipid (L). *Source:* Erasmus 1975.

salt of calcium, which originate in the cisternae of the GER. A count of cells in the various developmental stages in mature *S. mansoni* females was made by Erasmus and Popiel (1980), who estimated 39% S1, 6% S2, 30% S3, and 25% S4.

In *S. mansoni* cultured in vitro from cercariae to adults, Irie et al. (1983) found the vitelline cells to be essentially normal, demonstrating that specific host influences are not necessary for their development.

A detailed comparison of vitelline cell development in *S. mansoni, S. haematobium, S. japonicum*, and *S. mattheei* was undertaken by Erasmus et al. (1982). Similarities are great, particularly among the three African species. In *S. japonicum* the stage 4 (mature) vitelline cells contain somewhat different-appearing vitelline droplets, the lipid droplets are not associated with glycogen, irregularly-shaped bodies resembling ribosomal whorls are found, and calcareous granules are absent. These differences may be associated with the far greater rate of egg production in *S. japonicum*.

In *S. douthitti*, Nez and Short (1957) have provided a sketch of vitelline cell development. Young cells within the vitelline gland are relatively small, averaging around 9.5 μm in diameter. Many mitotic figures are evident. When the cells enter the vitelline duct they have accumulated considerable granular material and are around 17 μm in diameter. As the cells move forward in the duct, the granules, formerly uniformly dispersed, become arrayed near the cell membrane. By the time the cells enter the ootype the granules have been released and lie free in the duct of the ootype.

Lipid droplets in the mature vitelline cell have been identified as neutral lipid enclosed in a phospholipid-containing membrane (Doughty 1975).

Physiology and Function. The high demands of egg production dictate an active metabolic role for the vitelline gland. The intimate relationship between the nutrient-providing intestinal ceca and the vitelline gland was considered by Senft (1969), who emphasized the thin boundaries between these organs. Nollen et al. (1976) demonstrated that tritiated thymidine was incorporated into vitelline cells within 2 hours after incubation, and found some labeled cells in the vitelline duct of *S. japonicum* 1 day, and of *S. mansoni* 2 days after labeling. Avid uptake of tritiated thymidine specifically into stage 1 vitelline cells of mature paired female *S. mansoni* was demonstrated by Den Hollander and Erasmus (1984). Assuming that each female *S. mansoni* produces 300 eggs per day, and there are 38 vitelline cells per egg, Erasmus and Davies (1979) calculated that the vitelline gland in this species must release about 11,000 mature cells daily. In *S. japonicum* this figure is higher perhaps by a factor of ten. It was estimated by Becker (1977) that a female *S. mansoni* converts nearly her own body weight into eggs every day, although Grant and Senft (1971) calculated that the same worm producing 150 eggs daily would discharge only about 10% of her weight.

The uptake of tyrosine was studied by Erasmus (1975), who injected the labeled amino acid into the peritoneal cavity of infected mice. Rapid uptake into vitelline cells was noted, with specific localization in the vitelline droplets and endoplasmic reticulum. On the other hand, proline, present in high concentration in the egg, was not taken up heavily by the vitelline gland of *S. mansoni*, although the ovary and Mehlis' gland did accumulate the label (Senft 1968). Phenol oxidase activity was induced in female *S. mansoni* by incubation in serum-containing medium. It was suggested that increased enzyme activity was a result of accumulation of an enzyme with a high rate of turnover rather than by activation of a latent enzyme (Seed et al. 1978). In adult female *S. japonicum*, Wang et al. (1986) demonstrated by isoelectric focusing the presence of several isoenzymes of phenol oxidase. These authors tested several substrates, the best of which was found to be DOPA.

According to Doughty (1975), immature vitelline cells synthesize shell precursor material and then switch to production of lipid globules, which provide nourishment for the miracidial embryo within the eggshell. It was believed by Gönnert (1955a), however, that lipid droplets represent an end product of metabolism of the dying vitelline cell, because the extraembryonal lipid within the eggshell is not depleted during development of the embryo.

Effect of Drugs. Because of its importance in egg production and its high biosynthetic turnover, many workers have studied the effects of antischistosomal drugs on the vitelline gland. Among the numerous reports of drug effects on the vitelline gland the following may be noted:

Bang and Hairston (1946) showed that fuadin caused almost complete destruction, with loose vitelline cells found in the uterus of *S. japonicum*. Erasmus (1975b) found that astiban had a highly selective action on the vitelline cells of *S. mansoni*, destroying cells at stages 2 and 3. Mature cells were less affected, and stage 1 cells not at all so that they were able to resume development upon withdrawal of the drug. Antimony was accumulated specifically by stage 4 cells, particularly within the vitelline droplets and ribosomal complexes.

Erasmus and Popiel (1980) tested lucanthone and hycanthone, demonstrating a specific action against stage 1 cells, which eventually disappeared so that the lobules gradually became filled with mature vitelline cells, and replenishment did not occur upon withdrawal of the drugs. Similar results were reported by Popiel and Erasmus (1979).

Gönnert (1947) studied the effect of miracil, which destroyed maturing cells in the testes, ovaries, and vitelline glands.

Jackson et al. (1968) recorded the ability of two unrelated substances, ethylenedimethanesulphonate (EDS) and hexamethylphosphoramide to inhibit reproduction of *S. mansoni* in mice. Worms from treated animals differed from those in untreated controls primarily in the gonads, which were small in both sexes, and in the atrophic vitelline glands.

Moczon and Swiderski (1983) used *S. haematobium* to observe effects of niridazole. As early as the second day after treatment, a decrease was seen in early vitelline cells and ultrastructural changes noted, including alteration of intracellular vitelline droplets. By 5 days further degenerative changes were noted.

Popiel and Erasmus (1981b) also employed niridazole, but on *S. mansoni*. Progressive deterioration of the vitelline gland began at the posterior end of the worm, resulting in degeneration of mature vitelline cells and eventual deletion of entire lobules.

Procarbazine had dramatic effects on testes of male *S. mansoni* but did far less damage to ovaries of females (Basch and Clemens 1989).

The Mehlis' Gland

This rather mysterious organ extends posteriorly between the anterior limits of the ootype and the ovary, as shown diagrammatically in Fig. 3–5. The Mehlis' gland surrounds the gut, parts of the osmoregulatory system, and parts of the vitelline and oviducts and their connectives, is shot through with nerve and muscle fibers, and is one of the most complex parts of the entire organism. As a key part of the eggmaking apparatus, it is deserving of concentrated attention. Secretion of the Mehlis' gland was described by Gönnert (1955b) even in unisexual *S. mansoni*.

Early parasitologists considered that this was the "shell gland" that produced the hard outer layer, but it is now certain that the shell is made from coalesced vitelline droplets. Functions proposed for the Mehlis' gland include (1) lubrication of the uterus for the passage of eggs; (2) activation of spermatozoa; (3) release of shell globules from the vitelline cells; (4) control or initiation of the quinone tanning process; and (5) providing a membrane that serves as a template upon which the shell droplets accumulate. The bulk of evidence suggests the last alternative (Smyth and Halton 1983).

In *S. japonicum* Ho and Yang (1974) described the Mehlis' gland as consisting of a single type of unicellular gland cells. Their secretions, strongly PAS positive even after enzyme treatment, were thought to help in coalescence of vitelline granules and formation of the eggshell. No quantitative analysis of Mehlis' gland secretion has been done in schistosomes, although this has been studied in some other trematodes such as *Fasciola* (Clegg and Morgan 1966).

Ultrastructurally a syncytial, tegumental type tissue is found. The apical cytoplasm is extensively subdivided into long processes that pass through to the lumen of the ootype. The cytoplasm is filled with sacs of distended RER and secretory granules, which when mature pass below the muscular layer of the ootype where they accumulate in large numbers (Spence and Silk 1971a). Here they apparently dissolve and are discharged across the plasma membrane into the ootype lumen. A detailed, very well illustrated study of this organ was made by Spence and Silk (1971b).

The Ducts of the Female Reproductive System

Seminal Receptacle and Oviduct. Some authors, such as Faust et al. (1934), have mentioned a lateral cecum or blind pouch in the proximal oviduct acting as a seminal receptacle in *S. mansoni*. Gönnert (1955a) described an ampulla just at the origin of the duct, having strong sphincter muscles and large enough for a single egg, and a similar structure exists in *S. japonicum*

(Ho and Yang 1974). According to Severinghaus (1928), some individuals of *S. japonicum* possess a distinct stalked saccular seminal receptacle just at the proximal end of the oviduct, and others merely have a dilated lumen of the main duct in the same region. At least in *S. mansoni* and *S. japonicum* there is considerable variation in this structure, which seems to be specialized for storage of incoming sperm. In *S. bovis* the seminal receptacle consists of a caplike extension on the posterior end of the ovary, opening into the oviduct.

Control of the flow of cells and materials may be through a series of valves. Ho and Yang (1974) described valvelike structures consisting of 2 or 3 nuclei in *S. japonicum* at the junction of the ovarian ampulla and the proximal oviduct, the junction of the vitelline and oviducts, and in the lumens of the vitelline and ovovitelline ducts.

In *S. mansoni* the oviduct cytoplasm is syncytial in nature, containing mitochondria, Golgi zones, and a ribosomal component. Some secretory activity occurs in the ciliated seminal receptacle region. The duct is enclosed by a single layer of circular myofiber bundles, supported by a meshwork of sarcoplasmic processes that contain stores of glycogen, ribosomes, and mitochondria (Spence and Silk 1971a).

Erasmus (1973) reported that the characteristics of the cells forming the walls of the female reproductive ducts suggested that they may have a digestive function (Fig. 3–8).

Ootype. The most detailed study of the ootype was made by Gönnert (1955a), who compared this region in *S. mansoni, S. japonicum,* and *S. haematobium.* Where the common ovovitelline ducts open into the ootype, a valve made of 2 cells with large nuclei extends into the lumen. Muscular contraction and relaxation control the function of this sphincter, which seems to prevent backflow of egg materials but permits the upstream passage of sperm. Immediately downstream from the sphincter is an area 20 to 30 μm long and 10 μm wide into which the granular secretions of the Mehlis' gland empty through many elongated ducts. The main chamber of the ootype reflects the size and shape of the finished egg, whether ovoid or semilunar, including lateral or terminal spines if present. Its syncytial epithelium consists of serous apocrine gland cells with a rhythmic function correlated by Gönnert (1955a) with the cycle of egg production into four phases: resting, production, restitution, and extrusion. This detailed analysis, significant for an understanding of egg production, has apparently not been reinvestigated nor confirmed by other workers. Secretory granules are intermediate in nature between those of the ovi-and vitelline ducts, and the uterus (Spence and Silk 1971a). Ootype secretions were considered by Gönnert (1955b) of the greatest importance for confluence of vitelline granules and assembly of the eggshell.

Uterus. This duct extends from the anterior margin of the ootype, probably bearing a controlling sphinter at either end. It possesses a typical tegumental

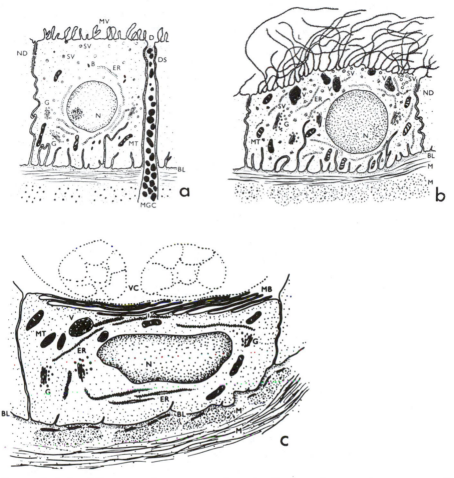

FIGURE 3–8 Semidiagrammatic representation of cells from the reproductive tract of female *S. mansoni*. a, Transverse section through the epithelium at the posterior end of the ootype. The luminal surface bears spatulate microvilli (MV) and receives the ducts (D) from the cells of the Mehlis' gland (MGC). The duct wall is attached to the ootype cells by septate desmosomes (DS), whereas adjacent ootype cells are in contact via nonseptate desmosomes (ND). The basal cell membrane is elevated into the cell cytoplasm and rests on a granular basement layer (BL). The cell cytoplasm contains a nucleus (N), numerous Golgi complexes (G), mitochondria (MT), and small strands of granular endoplasmic reticulum (ER). Also present in the cytoplasm are small vesicles containing granular material (SV) and dense bodies (B). b, Transverse section through the nonciliated region of the oviduct. The oviducal cell bears lamellae (L) coated with a granular material. There are large, irregular vesicles (V) within the cytoplasm, and a muscular coat (M) beneath the basement layer. Other abbreviations as in (a). c, Longitudinal section through a cell from the wall of the vitelline duct. The luminal surface is elevated into stacks of membranes (MB), which are often compressed by vitelline cells (VC) in the lumen of the duct. Other abbreviations as in (a) and (b). *Source:* Erasmus 1973.

structure but the apical cytoplasm is found in a very broad band; luminal surface features vary along its length to the gonopore, where it everts to become continuous with the outer tegument (Spence and Silk 1971a).

The ducts of the female reproductive system of *S. mansoni* show regional differences in immunoreactivity to antisera against ecdysterone (Basch 1986a) and a variety of neuropeptides (Basch and Gupta 1988).

Typical uterine egg counts in schistosomes are shown in Table 3–5.

Oogenesis and Egg Formation

Because of its pivotal signficance in pathogenesis and epidemiology, the schistosome egg should be a prime target for pharmacological attack. However, surprisingly little is known about the development, structure and composition of these eggs. Extant reports are marked by a lack of descriptive and nomenclatorial uniformity: in particular the terms "egg," "ovum," "capsule," "oocyte," and "yolk" have been utilized in various ways by different authors in English, and the situation is further complicated by publications in other languages.

In his comprehensive review of oogenesis in the Digenea, Gresson (1964) discussed the cytological development of oocytes, the entry of spermatozoa, and details of fertilization but mentioned no reports on the Schistosomatidae. However, as early as 1928 Severinghaus had noted that maturation of oocytes was not completed within the ovary of *S. japonicum*. Reproductive cytology was discussed in some detail for *S. mansoni* by Niyamasena (1940), who showed that it is the primary oocyte that leaves the ovary and is later penetrated by the incoming sperm, and indicated that actual fertilization and subsequent cleavage and development occur within the eggshell. The most detailed observations on these phenomena in schistosomes are for *Schistosomatium douthitti:* Short (1952a) reported that almost every intrauterine oocyte in bisexual infections contained a single spermatozoon in its cytoplasm. Nez and Short (1957) found penetration by a sperm cell to occur in the proximal oviduct soon after discharge of the primary oocyte from the ovary. Further maturation and actual fertilization (i.e., nuclear fusion) occur only after the eggs have been discharged from the female worm and deposited in host tissues.

Assembly of the Egg

The complex process of egg production in trematodes has been studied since the 1880s when their life cycles and general biology were first elucidated. Nevertheless many mysteries remain, including the nature and function of the Mehlis' gland complex and of the other units of the female reproductive system, particularly the ootype epithelium; the precise biochemistry of shell formation; and the mechanisms for control and integration of this highly sophisticated series of procedures.

The structure commonly termed the egg is assembled from the oocyte (normally with contained spermatozoon) plus several mature vitelline cells, with the aid of glandular secretions that are until now poorly characterized (Fig. 3–9). In *S. japonicum* the number of vitelline cells ranges from 16 to 30 and is usually about 20 (Ho and Yang 1974); in *S. mansoni* it is normally about 30 to 40 (Gönnert 1955b).

Eggshell formation in helminths has been the subject of a comprehensive comparative review by Wharton (1983). As a general rule in trematodes, eggshell formation begins in the ootype, which is more or less filled with a mucoid secretion before the entry of the bolus of egg-forming materials. This secretion becomes layered on the wall of the ootype and forms a thin coat over the cells destined to make up the egg (Ebrahimzadeh 1966). In *Fasciola hepatica* the Mehlis' gland secretion has been identified as lipoprotein by Irwin and Threadgold (1972), whose excellent illustrations should be consulted. Vitelline granules are freed from the vitelline cells, perhaps with mechanical assistance from squeezing through the narrow sphincter proximal to the ootype, and migrate to the exterior of the package. Here they begin to coalesce starting from the outside so that the shell gradually thickens inward. In *S. mansoni* eggshell formation has been described in detail by Gönnert (1955b), who believed that secretion of the ootype epithelium is of the greatest importance for the confluence of vitelline granules to form the membranous shell.

The process of eggshell formation appears to be a quinone tanning system, reviewed in detail by Nollen (1971), Wharton (1983), and Smyth and Halton (1983). The shell proper is composed of sclerotin, a tanned protein of wide occurrence in the animal kingdom. This material is derived from proteinaceous and phenolic precursors in the vitelline granules, whose presence has usually been determined by a battery of histochemical tests (but Smyth and Halton 1983 have advised caution in interpretation). Phenol oxidase activity has been reported in *S. mansoni* (Bennett et al. 1978, Seed et al. 1978) and *S. japonicum* (Ho and Yang 1974) and the tanning process appears to be similar in *S. douthitti* (Nez and Short 1957). The eggshell may continue to thicken and grow in size after oviposition. A full elucidation of the biochemistry of eggshell polymerization is much needed, as this represents a useful point of attack against the parasites.

Aberrations. Various errors and aberrations of egg formation have been reported. In *S. mansoni* oddly-shaped eggs have been correlated with an abnormally formed ootype (Bruijning 1968b), which forms the eggs in a mold-like process (Fig. 3–10). Worms may discharge loose oocytes, vitelline cells, and "vitelline conglomerates" which are antigenic (Kelley and von Lichtenberg 1970). The significance of host nutrition was emphasized by Krakower et al. (1944), who reported defective eggshell formation in *S. mansoni* maintained in vitamin C-deficient, but not in vitamin C-supplemented guinea pigs. Paraense (1949) observed and illustrated a variety of misshapen and oddly formed eggs and vitelline conglomerates produced by unisexual female *S. mansoni* of a strain collected in Manguinhos, Rio de Janeiro, Brazil,

TABLE 3–5 Uterine Egg Counts, Female Schistosomes[a]

Species/Origin	Host	Natural/Experimental	Age, Days	Egg Count Range	Egg Count Mean	Comment	Reference Note
S. mansoni							
Puerto Rico	Mouse	E	40	0–1	.84–.90		1
Puerto Rico	Monkey	E	55	0–25	Usually 1	322/346 had 1	2
Liberia	Mouse	E	40	0–1	.51–.64		3
Egypt	Monkey	E		0–3	Usually 1		4
Egypt	Mouse ?	E		0–4	Usually 1		5
S. rodhaini							
Kenya	Hamster	E			1		6
Uganda	Various				1		7
S. haematobium							
Sudan	Hamster	E		–113	34	Max @ 120d	8
S. Africa	Hamster	E		–78	22	Max @ 130d	
Egypt	Hamster	E	120		121	Maximum	9
Egypt	Hamster	E	250		34		
Egypt	Hamster	E	90		146	Maximum	
Egypt	Hamster	E	250		39		
Egypt	Human	N	?	13–55	16–43	6 groups	10
Egypt	Human	N	?		36–83	Age dependent	11
Egypt	Monkey	E		to 50			12
Iran	Hamster	E	to 170	Varies	21–57	See data	13
Iran	Hamster	E	to 150	Varies	7–53	See data	14
Nigeria	Hamster	E	to 171	Varies	19–31	See data	15
Durban	Hamster	E	to 170	Varies	14–35	See data	16
Mauritius	Hamster	E	to 150	Varies	7–56	See data	17
Ghana-K	Hamster	E	to 150	Varies	10–42	See data	18
Ghana-P	Hamster	E	to 150	Varies	12–34	See data	19

Species / Locality	Host						Ref.
S. bovis							
Corsica	Mouse	E					20
Sudan	Cattle	N	?	3–39	20–30		21
Israel	Mouse	E		13–57	14.1		22
*	*	*	51	3–65	17.5	47 worms	23
*	*	*	57				24
S. intercalatum							
Cameroon	Hamster	E	to 130	Varies	12–60	See data	25
S. leiperi							
Zambia	Hamster	E	to 100	Varies	6–17	See data	26
S. indicum							
India	Pig	N	?	Always 1	1		27
S. margrebowiei							
Botswana	Hamster	E	to 60	Varies	30–200		28
S. mattheei							
Transvaal	Hamster	E	to 130	Varies	5–42		29
Transvaal	Cats	E		14–56			30
*	*	–	–	70–95			31
S. Africa	Sheep	E		0–72	23	Large female	32
S. nasalis							
India	Buffalo	N	?	1–5	Usually 1		33
S. japonicum							
Japan	Rabbit	E			91.5	109 worms	34
	Mouse	E			88.7	12 worms	
	Guinea Pig	E			86.9	129 worms	
	Rat	E	30		13.8		
China	Mouse	E			ca. 30	134 worms	35
China	Guinea Pig	E			124.5	61 worms	36
China	Mouse	E	28	14–54	72.3		37
			40	0–202	74.9		
			50	62–215	122.4		

TABLE 3–5 Uterine Egg Counts, Female Schistosomes[a] (continued)

Species/ Origin	Host	Natural/ Experimental	Age, Days	Egg Count Range	Egg Count Mean	Comment	Reference Note
Philippines	Mouse	E	25–46	0–132	46.3		38
S. mekongi				120–130			39
Trichobilharzia ocellata							
Canada	Ducks	E	9	0–1	Usually 0	40 worms	40
Austrobilharzia variglandis							
Hawaii	Birds	N			Usually 1		41
Microbilharzia variglandis							
Rhode Island	Birds				1		42

[a]*See also* Loker, 1983.

*not specified

Reference notes: 1, Bruckner and Schiller 1974; 2, Faust et al. 1934; 3, Bruckner and Schiller 1974; 4, Fairley 1920; 5, Leiper 1918; 6, Saoud 1966b; 7, Fripp 1967; 8, James and Webbe 1975; 9, James and Webbe 1973; 10, Cheever et al. 1975; 11, Cheever et al. 1977; 12, Fairley 1920; 13, Webbe and James 1971b; 14, Wright and Knowles 1972; 15, Webbe and James 1971b; 16, Wright and Bennett 1967a, b; 18,19, Wright and Knowles 1972; 20, Brumpt 1930a; 21, Dinnik and Dinnik 1965; 22, Lengy 1962b; 23,24, Brumpt 1930c; 25, Wright et al. 1972; 26, Southgate et al. 1981; 27, Sinha and Strivastava 1956; 28, Southgate and Knowles 1977; 29, Wright et al. 1972; 30, Dinnik and Dinnik 1965; 31, Brumpt 1930c; 32, van Rensburg and van Wyk 1981; 33, Dutt 1967; 34, Ito 1954; 35, Cort 1921; 36, Vogel and Minning 1947; 37, Hsü et al. 1958; 38, Pesigan et al. 1958; 39, Kitikoon 1980; 40, Bourns et al. 1973; 41, Chu and Cutress 1954; 42, Stunkard and Hinchcliffe 1952.

122

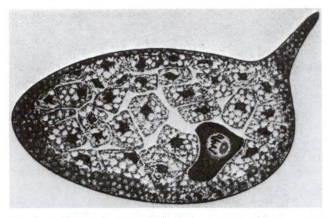

FIGURE 3–9 Developing egg from the ootype of *S. mansoni*. Vitelline droplets around the periphery, and "empty" vitelline cells around the ovum with chromosomes. *Source:* Gönnert 1955b.

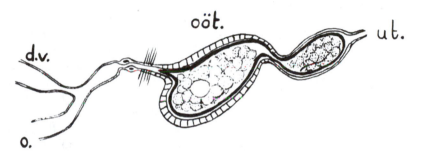

FIGURE 3–10 Abnormally shaped egg of *S. mansoni* and its method of formation within a distorted ootype. d.v., vitelline duct; o, oviduct; oöt, ootype; ut, uterus. *Source:* Bruijning 1968.

which resemble the egglike objects arising from *S. mansoni* grown from cercariae in vitro by Basch (1981b).

Drug Effects. Most antischistosomal chemotherapy seems to affect the delicate working of the female reproductive system and disrupt the process of egg assembly (Monteiro et al. 1949). Among the drugs reported to have adverse effects on egg and eggshell formation are: disulfiram (Bennett and Gianutsos 1978), tris (p-aminophenyl) carbonium salts (Bueding et al. 1967), nicarbazin (Campbell and Cuckler 1967), nine thiourea compounds (Ho and Yang 1974), and thiosinamine (Machado et al. 1970, Popiel and Erasmus 1981c). A variety of specific effects was noted in these and other studies of inhibitory drugs, including depression of phenol oxidase activity, inhibition of tanning, and production of distorted eggs devoid of normal contents.

Direction and Coordination of Egg Production

The direction and coordination of the machinery of egg production has not been specifically described in schistosomes, and work on this difficult subject is needed. These processes may be similar in principle to the system described by Gönnert (1962) for *Fasciola hepatica* (freely translated):

> The oogentop is . . . complicated in structure. Therefore there must be a coordinating system, whose function it is to regulate the flow and the proper sequence of individual functions. To this question there are as yet no data available. . . . In my investigations I have seen nerve cells in the region of the Mehlis' gland. Precise examination of serial sections revealed the existence of two groups of nerve cells, which are called Plexus I and Plexus II. Plexus I consists of usually 4 and at most 5 nerve cells which lie in the region of the ovovitelline duct and presumably coordinate the functions of the oviduct, vitelline duct, ovovitelline duct, and glands of the upper ootype and Mehlis' gland. . . . Plexus II consists of only 2 nerve cells. These lie close to the upper uterus. One of these is found very near the ootype valve and appears to be connected to it by a process. The second cell lies further towards the uterus. I presume that Plexus II innervates the lower ootype and the ootype valve as well as the upper uterus section, thus the region of eggshell formation.

Just how such nerve plexi function remains a mystery. It is possible that sensory nerve endings of the stretch receptor type respond to the presence of materials in the reproductive ducts with neural feedback loops through the effector nerves to control the function of glands and sphincters. Despite the report and review of Dei-Cas et al. (1980) complete with diagram of the neural ganglia of *S. mansoni* (Figure 1–4), little information is available on schistosome neuroanatomy. The few detailed studies made with the electron microscope (Silk and Spence 1969b; Reissig 1970; Dei-Cas et al. 1980) agree in describing a variety of electron dense granules, some of which were considered to be neurosecretory by all authors. No function was ascribed to these intraneuronal granules and no study has been made of their association with developmental states nor with components of the reproductive system. By analogy with other invertebrates, neurosecretory control of reproductive functions, through peptide neurohormones, appears a reasonable premise in schistosomes. It is also possible that other internal hormones such as ecdysones play a role in reproductive regulation (Nirde et al. 1983, 1984) but at the moment this is speculative.

In vitro studies can be useful in defining conditions needed for proper egg production and development. It has been reported (Basch 1981) that paired *S. mansoni* cultured in vitro from cercariae to adults will produce imperfect, infertile eggs. This was found also in cultured *S. douthitti* (Basch and O'Toole 1982), in which both mixed sex and unisexual female cultures yielded eggs, thereby resembling the situation in vivo. In an ultrastructural study of the nature of the imperfections in "cultured" eggs of *S. mansoni*, Irie et al. (1983) reported degeneration in oocytes, as mentioned above. Survival and egg production by ex-vivo adult *S. mansoni* in nine chemically defined media was studied by Newport and Weller (1982), who emphasized the role of lipids. Basal media supplemented with delipidated bovine serum albumin plus a fatty acid mixture rich in stearic acid provided results comparable to those obtained with media supplemented with 8% newborn calf serum. It was found by Schiller et al. (1975) that adult *S. mansoni* could be maintained as well in media in the presence or absence of oxygen; however, almost no eggs were produced under anaerobic conditions in vitro.

Internal chemical communication within individual flatworms has been reviewed by Basch (1986b) but little information is available about the neuronal or hormonal processes governing reproductive processes in trematodes in general or in schistosomes in particular.

Egg Laying Behavior

The precise manner in which females oviposit is still a matter of some speculation. The suggestion by Lutz (1919) that the female *S. mansoni* passes through the venule wall to place eggs in the tissues can probably be discounted, as can the influence of external temperature (Maciel 1924) or barometric pressure (Maciel 1926) on egg-laying of schistosomes. It was suggested by Manson-Bahr and Fairley (1920) that "when desirous of depositing ova" the female *S. mansoni* or *S. haematobium* protrudes herself from the male, which is held stationary within the blood vessel by the tuberculated tegument. The female then creeps into very fine venules and deposits eggs, whose spine is directed against the blood flow to help anchor it in the vessel (Fig. 3–11). Withdrawal of the female to return to the larger vessel where she left her male partner causes the elastic vessel wall to contract and hold the freshly deposited egg. A series of small backwards movements results in deposition of a series of linearly placed ova within the venule. Females of both species, of Egyptian origin, were often found singly in monkeys by Manson-Bahr and Fairley (1920) and Fairley (1920). Similarly, Faust et al. (1934) reported from Puerto Rico that only a small proportion of their worms was found coupled, and therefore mating was a relatively temporary phenomenon for females of *S. mansoni*. It was found likely by Faust et al. that

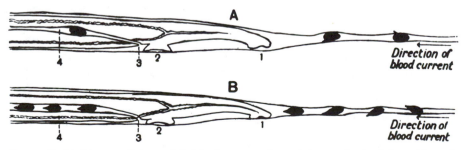

FIGURE 3–11 Manson-Bahr and Fairley's conception of the method of deposition of eggs into the lumen of a small venule, and their passage through the vessel wall. *Source:* Manson-Bahr and Fairley 1920.

the egg was discharged only when the female elongates into a small venule, the constriction of the anterior part of her body setting in motion the uterine contractions needed to force the egg outward. In both the Egyptian and Puerto Rican *S. mansoni,* female worms were always found with their heads directed toward the terminus of the venules, but Faust et al. reported that 12% of males faced in the other direction. The male schistosome, meanwhile, was considered by Brumpt (1930c) to act as a sort of piston to temporarily reverse the blood flow in order to inject or project the eggs into the most distal venules. This idea was rejected by Faust et al. (1934), although it was conceded that the male, by retarding blood flow, could cause expansion of the terminal venules.

Whether the female schistosome leaves the male is an unresolved question, the early history of which has been reviewed in detail by Brumpt (1930c). Once again, observations have been made on only a few species and nothing seems to be known, for example, from any of the avian schistosomes. Many observers have found essentially all adult worms of *Schistosoma,* numbers permitting, to be paired within the blood vessels. For example, Pellegrino and Coelho (1978) have stated that at the time of oviposition the female is held in the gynecophoric canal of the male where she deposits an egg, retreats a bit, then lays another egg, and so on until a beadlike row of immature eggs is produced. Brumpt (1949) has also reported that female schistosomes are almost always paired while advancing in the submucosal venules, and Faust and Meleney (1924) reported a similar observation in *S. japonicum* from China. In heavy experimental infections worms were found seldom to lie alone, but were usually mated and occurred in bundles of 6 to 10, even in the minor radicles of the mesenteric vein. In my own laboratory, *S. mansoni* adults in balanced bisexual infections are almost always found paired when dissected from the mesenteric veins of a mouse or hamster sacrificed by rapid cervical dislocation. The method of sacrifice of the host, use of various drugs or anesthetics, or technique of perfusion may introduce artifacts to influence the degree of pairing observed.

The eggs of *S. japonicum* were found by Pesigan et al. (1958) to occur in densely packed clusters of about 116 (70 to 120) within the gut wall of experimental animals. The eggs in any cluster were all at the same stage of development, suggesting that they were all laid at one time by one female, "which probably empties her uterus in the act." The number of egg-layings per day rose from less than 1 during the first 3 days of maturity to nearly 12 after 2 weeks, with a corresponding increase in total egg production.

The Schistosome Egg

From an anthropocentric viewpoint the schistosome egg deserves special attention. It is the primary inducer of pathogenesis in the host, and also the agent of transmission.

In human infections, the egg remains within the host tissues for long periods: roughly 6 days for eggs that will exit the host, and for months or perhaps years in the case of those eggs trapped in the liver or other tissues. The Hoeppli phenomenon, the induction of granulomatous tissue host responses, and the careers of a phalanx of immunopathologists are based on the leakage of a variety of soluble egg antigens through micropores in the eggshell. Also, particularly in *S. japonicum* infections, the egg may be carried through the circulatory system to remote sites such as the brain.

In human infections, the egg is the confirmatory unit of diagnosis, and is usually the only stage of the parasite to be seen except where miracidial hatching tests are done. Where the species of schistosome in humans is not obvious (as it would be in Brazil or the Philippines, for example) specific identification of the parasite generally rests on egg morphology. The same is true for veterinary and avian schistosomes: for example, Appleton (1982) used egg morphology to distinguish infections of *Gigantobilharzia, Austrobilharzia,* and *Trichobilharzia* in South African wild birds.

In Africa, with its complex of species potentially infective to humans, the size and shape of the eggs represents the diagnostic gold standard. At the time when schistosome species were poorly differentiated, it was believed by Looss that terminal-spined eggs were the normal product of impregnated females of *S. haematobium,* while lateral-spined eggs were produced parthenogenetically where males were not developed. Manson felt that the two types of eggs emanate from zoologically distinct parasites, a viewpoint finally confirmed by Leiper (1916). Although it is important to study variations in dimensions of eggs of each species, as was done for example by Schwetz (1953) for *S. haematobium* and *S. intercalatum,* there may still be intermediate specimens that make specific determination difficult. Egg measurements may vary with several factors: (1) there is generally an increase in size between intrauterine eggs and those deposited in the tissues or found in stools; (2) host factors including chemotherapy may affect the eggs; (3) the species of host may influence egg size, as Hsü and Hsü (1958a) pointed out for *S. japonicum;* and (4) the species of host may influence egg shape. Lengy (1962a) showed that although eggs of *S. bovis* from mouse, hamster, and sheep were similar, those from the guinea pig were long, narrow, and asymmetrical. Convergent evolution may complicate the interpretation of egg morphology. The egg of *S. sinensium,* a member of the *S. japonicum* complex, is remarkably similar to that of *S. mansoni* and may easily be confused with that species (Greer et al. 1989). A table of dimensions of mammalian schistosome eggs is presented by Loker (1983).

Numbers Produced; Fecundity

The number of eggs produced by a female worm is significant both for pathogenesis and transmission. Trematodes have a high reproductive potential: Boray (quoted by Whitfield and Evans 1983) has estimated the fecundity of

Fasciola hepatica at 25,000 eggs per worm per day. While the known egg production of schistosomes does not approach that number, females are nevertheless highly active reproductively. Grant and Senft (1971) estimated that at an average daily production of only 150 eggs, the female *S. mansoni* discharges 10% of her weight as eggs every day. Sturrock's (1966) estimate that a *S. mansoni* egg weighs 1.7×10^{-7}g and a mature female worm weighs 2.15×10^{-4}g is roughly in the same range. Various estimates and counts of schistosome fecundity are shown on Table 3–6.

The determination of daily egg production by schistosomes within a host is a very imprecise excercise. Therefore it is not surprising, as Sturrock (1966) found, that estimates of daily egg production by a female *S. mansoni* have ranged from 2.1 to 35,000! As a general guide, a female *S. haematobium* produces roughly one egg each 10 minutes; *S. mansoni,* each 5 minutes, and *S. japonicum* each minute.

Numerous factors influence the number of eggs produced by a female schistosome, and except for closely controlled experiments it is not possible to disaggregate the intrinsic characteristics of the worms from environmental and host factors. Host genetic influence on fecundity was studied by Jones et al. (1989) in two strains of mice, NIH/01a and CBA/Ca. The timing of peak excretion differed in the two strains, as did the distribution of eggs in the tissues. The NIH/O1a mice had more eggs in the liver, the CBA/Ca mice had more in the lungs and gut.

In Ghana, Onori (1962) followed the daily urinary egg output in 200 boys aged 5 to 15 infected with *S. haematobium.* Urines were collected at 8 and 11 A.M. and at 2 P.M. Most eggs were found in the 2 P.M. specimen and the fewest at 8 A.M. The relationship between average egg output and prevalence of infection in each site was, as expected, direct and significant.

Factors Controlling Egg Output

Worm-worm Interactions; Density Dependence of Fecundity. In a carefully documented study of *S. mansoni* in baboons, Damian and Chapman (1983) estimated a mean output of $1,107 \pm 258$ (SE) eggs per female per day. As this figure was similar in animals with a 10-fold difference in worm burden it was considered that crowding does not affect fecundity within the range observed. On the other hand, Jones et al. (1989) reported for *S. mansoni* in mice that "there is significant evidence for a density dependent reduction in fecundity of worms in more heavily infected animals." For human infections, the same data were interpreted in opposite ways by Medley and Anderson (1985) who claimed a density-dependent reduction in egg output, and by Cheever (1986) who found no such effect.

Damian and Chapman (1983) reported a reduction in fecundity of *S. mansoni* in baboons in cases where the sex ratio was significantly imbalanced. However, Cheever (1988) could demonstrate no significant effect on fecundity of *S. mansoni* in rhesus monkeys, or on the number of *S. mansoni* eggs in humans or mice, owing to imbalances between the numbers of male and female worms present in the hosts. The number of S. *japonicum* eggs in the tissues of rabbits was also said to be unaffected by any disproportion between worm genders.

Such statistical interpretations of presumed biological effects tend to be relatively esoteric; however, the underlying interactions, if substantiated, must be explainable on the basis of physiologic mechanisms.

Host Factors. The role of host immunity in worm fecundity is not clear, and statistical interpretations are confounded by increasing worm age and other factors. Nevertheless this is an important field of inquiry, particular in view of efforts to develop an antischistosomal vaccine.

Immunological aspects of resistance to schistosomes in mice have been reviewed extensively by Dean (1983): following a single infection, some mice have shown a reduction and complete cessation of egg excretion, while in others excretion rates remained stable for long periods. Fecal or urinary egg output is a measure neither of the number of female worms or worm pairs present, nor of the rate of oviposition, because a substantial percentage of eggs is retained or destroyed within the host tissues. Although the interpretation of fecal egg counts is thus made more difficult, sudden rapid declines in host egg output are nevertheless significant. For example, Cheever and Powers (1972) found that eggs in the feces of rhesus monkeys infected with S. *mansoni* declined rapidly after 80 days post infection. Necropsies showed that the decrease in the numbers of eggs was proportionally greater than the decrease in worm numbers, and that eggs were not retained in tissues in greater numbers. It was concluded that egg production by surviving worms is suppressed in prolonged infections. A similar phenomenon had been noted earlier by Standen (1949). The situation is complicated, as Amin and Nelson (1969) determined, by the interrelationships of successive schistosome infections with different species: a prior exposure to S. *mattheei* was found to reduce the mean tissue egg count of a subsequent S. *mansoni* infection by 74%.

Surgical transfer of S. *mansoni* worm pairs from infected to naive nonhuman primate hosts does not lead to an increase in fecundity, failing to support the hypothesis that reduced fecundity in nonhuman primate hosts is a reversible result of host immunity (Damian et al. 1986).

Through an innovative approach, Wu et al. (1985) have looked at factors in portal serum that may promote egg production in schistosomes. Incubating ex-vivo S. *mansoni* pairs in medium supplemented with fetal calf serum, or rat or human serum from peripheral or portal vein blood, they found much higher egg production when the portal serum was used. Fractionation of rabbit portal vein blood serum by ultrafiltration showed the 10 to 50 kilodalton fraction to be most stimulatory of egg production.

Prepatent periods (i.e., the interval between infection and appearance of eggs in the excreta) have been tabulated by Loker (1983) for many species of schistosomes, and need not be repeated here. It is easier to determine the time of appearance of the first eggs in stool or urine of experimental animals

TABLE 3–6 Daily Egg Production, Female Schistosomes[a]

Species/Origin	Host	Natural/Experimental	Age, Days	EPFPD	Comment	Reference Note
S. mansoni						
Puerto Rico	Mouse	Exp	—	157–191	Variable	1
	Rhesus	Exp	p+7–8	Several 00	183, 239	2
	Rhesus	Exp		211	190–412	3
	Hamster	Exp		291	Control	4
	Mouse	Exp	40–54	225	Normal serum injection	5
	Mouse	Exp	40–54	274	Immune serum injection	
	Mouse	Exp	40–54	202		
Brazil	Monkey	Exp		529–658	Age varied	6
	Mouse	Exp	43–65	257		7
	Mouse	Exp	55–236	ca. 300	1 worm pair/mouse	8
Guadeloupe	*Rattus*	Nat	Unk	6, 70, 117	Wild rats	9
Egypt	Mouse	Exp	p+3wk	362	205–469	10
	Mouse	Exp	p+6wk	367	276–511	11
	Mouse	Exp		265		12
Kenya	Baboon	Exp		1,017	±258(SE)	13
S. haematobium						
Egypt	Hamster	Exp	250	215		14
Tanzania	Hamster	Exp	250	173		
Sudan	Hamster	Exp	250	130		15
S. Africa	Hamster	Exp	250	127		
Iran	Hamster	Exp	75–170	239		16
Nigeria	Hamster	Exp	75–150	185		
Durban	Hamster	Exp		105	28–281	17

Location	Host	Exp	EPFPD	p	Ref
Ghana-P	Hamster	Exp	160		18
Ghana-K	Hamster	Exp	159		
Iran	Hamster	Exp	177		
Mauritius	Hamster	Exp	189		
S. mattheei					
S. Africa	Calf	Exp	640–3,285	begin @6–7 wk	19
Transvaal	Hamster	Exp	148	99–268	20
S. haematobium X S. mattheei					
S. Africa	Hamster	Exp	269	230–316	21
S. intercalatum					
Cameroon	Hamster	Exp	207	166–311	22
S. bovis					
—	Mouse	Exp	60	begin @60 d — 79	23
	Mouse	Exp	110	87	
S. japonicum					
Taiwan	Hamster	Exp	3,500	2,984–4,350	24
China	Mouse	Exp	1,322	28 wk	25
China	Mouse	Exp	35–98	25 d	26
	Mouse	Exp	664	34 d	
	Mouse	Exp	2,092	44 d	
	Mouse	Exp	370	68 d	
Philippines	Mouse	Exp	100–1,400		27

aSee also Loker, 1983, Table 3, Pp. 350–51.

p = patency (first day of egg passage). EPFPD = Eggs per female per day.

Reference notes: 1, Brumpt 1930c; 2, Faust et al. 1934; 3, Nelson and Saoud 1966; 4, Moore and Sandground 1956; 5, Weinmann and Hunter 1961; 6, Cheever and Duvall 1974; 7, Kloetzel 1976b; 8, Pellegrino and Coelho 1978; 9, Imbert-Establet 1982; 10, Koura 1970; 11, Koura 1971a; 12, Koura 1971b, 72; 13, Damian and Chapman 1983; 14, James and Webbe 1973; 15, James and Webbe 1975; 16, Webbe and James 1971b; 17, Wright and Bennett 1967a, b; 18, Wright and Knowles 1972; 19, Lawrence JA 1973a; 20, Wright et al. 1972; 21, Wright and Ross 1980; 22, Wright et al. 1972; 23, Brumpt 1930a; 24, Moore and Sandground 1956; 25, Moloney and Webbe 1983; 26, Hsü et al. 1958; 27, Pesigan et al. 1958.

than to specify the day on which young female worms begin to produce and emit viable eggs. Such information is summarized on Table 3–6.

Egg Structure

The comprehensively illustrated description of S. mansoni eggs by Neill et al. (1988) should be consulted for ultrastructural detail.

The Shell

Surface Spines. Inatomi (1962) studied the ultrastructure of the shells of a number of trematodes: In *Paragonimus* and *Fasciola* the outer surface was quite smooth; in *Clonorchis, Opisthorchis,* and *Metagonimus* there was some form of projections. In *S. mansoni* and *S. japonicum* the surface was covered by small spines: those of the former were spinelike, about 0.08 × 0.3 μm, whereas those of *S. japonicum* were microvillilike, about 0.026 × 1 to 2 μm.

As part of an ultrastructural study of hepatic granulomas, Stenger et al. (1967) described the TEM appearance of *S. mansoni* eggs within mouse liver tissue. They found the shell to be covered by "innumerable" tiny spinelike external microspicules and multiple holes suggestive of fenestrations. Shortly thereafter Race et al. (1969) observed eggs deposited by female *S. mansoni* in vitro and demonstrated the presence of "microbarbs," which were thought to make the egg better able to cling to vascular endothelial lining and facilitate its penetration into tissues. Incubation in thioglycollic acid or pepsin failed to remove the microbarbs. Simultaneously, Hockley (1968) made a detailed TEM study of the egg shells of *S. mansoni* and *S. haematobium*. On *S. mansoni* eggs the small spines averaged about 0.05 μm wide at the base and ranged to 0.28 μm in length (agreeing almost exactly with the dimensions reported by Inatomi et al. 1970). On *S. haematobium* eggs the spines were about 0.22 μm long and somewhat less pointed. Whereas Hockley (1968) thought such rounding may have been due to wear, Schnitzer et al. (1971) examined *S. haematobium* eggs in utero in female worms and found the spines already somewhat rounded before deposition. Transverse sections of the spines revealed a dense inner core and outer part, made up of subunits and separated by a less dense area. The small spines, which formed a thin hair pile over the egg surface, were estimated at 375 million per egg by Hockley, increasing the surface area by three times over that of a smooth egg. They were thought possibly to trap egg secretions on the surface, thereby localizing their effect on adjacent cells. The spines could not readily be seen by SEM by Hockley (1968), but were visualized a decade later, using better methods, by Ford and Blankespoor (1979).

In the *S. japonicum* eggs from Japan described by Inatomi et al. (1970), large numbers of hollow microvillilike projections were seen, about 0.026 μm in diameter and about 1 μm in length, averaging about 900 per square micrometer of shell surface. Schnitzer et al. (1971), unaware of the work of

Inatomi, reported microspines 0.002 μm in diameter and 0.05 μm long on the surface of *S. japonicum* eggs (origin unstated). In *S. japonicum* eggs from China He et al. (1980) found about 600 microspicules per square micrometer, each about 0.02 by 0.06 μm in size. However, Sakamoto and Ishii (1976) found microvillilike chitinous projections "that very much resemble cobwebs" 0.05 to 0.08 μm in diameter and more than 3 μm in length on the eggs of *S. japonicum* (origin unstated; presumably from Japan). A netlike fibrous matrix visible by TEM was reported on *S. japonicum* eggs by Ford and Blankespoor (1979).

Thickness. Stenger et al. (1967) found the eggshell of *S. mansoni* to consist of a thin, very dense, apparently continuous inner layer and a thick, moderately dense outer layer. The major shell component was osmiophilic and became more dense after lead hydroxide staining. Neill et al. (1988) interpreted the *S. mansoni* eggshell as composed of three layers: the peripheral microspine layer 318 ± 14.9 nm thick, a middle layer with intermediate electron density 377.8 ± 24 nm thick that continues as the core of the surface microspines; and an extremely electron dense inner layer 59.5 ± 24 nm thick.

Inatomi et al. (1970) measured the thickness of the *S. japonicum* eggshell (Japanese isolate) at 0.43 to 1.09 (average, 0.68) μm, and that of the *S. mansoni* eggshell at 0.7 to 1.2 μm. The thickness of the *S. japonicum* eggshell from China was determined by He et al. (1980) to be about 0.4 μm, ranging from 0.26 to 0.48 μm. It is not known whether this difference is real or owing to variations in technique.

Eggs from *S. mansoni* grown completely in vitro consisted primarily of aggregated vitelline granules around which Basch and Basch (1982) found no shell at all.

Fenestrations. A fenestrated eggshell was described in *S. mansoni* eggs in mouse livers by Stenger et al. (1967), who stated that J.H. Smith and F. von Lichtenberg had independently observed the same structures. Greater detail was provided by Race et al. (1969), who demonstrated that discrete pores, located primarily in the anterior portion of the egg, traversed the shell directly. Each pore was lined by a trilaminar unit membrane. The average diameter of shell pores was 100 nm, and they were found to be lined by projections of the cytoplasmic membrane of the inner shell surface (Race et al. 1971). In the eggs of *S. japonicum* a series of micropores and a network of microcanals were found by Cao et al. (1982), also lined with a membrane continuous with the inner eggshell. In contradistinction to these pores, the eggshell fenestrations noted by Stenger et al. were considered to be artifactual (Race et al. 1971). Neill et al. (1988) found that "cribriform pores with serpiginous branching and anastomosing channels measuring 33.8 ± 3.7 nm in diameter (rather than direct inner-outer perforations) traverse the eggshell obliquely."

It is clear that the fenestrations can function as conduits for passage of materials in both directions through the shell. Erasmus (1973) suggested that nutrients necessary for miracidial development could be acquired from the surrounding host tissues through the "permeable" shell. Active uptake and utilization of tritiated thymidine and uridine, and of various labeled amino acids from culture medium were demonstrated independently at the same time by Stjernholm and Warren (1974). The egg of *S. mansoni* was shown to be an actively metabolizing entity with the ability to utilize exogenous materials. It was suggested also (Stenger et al. 1967) that the diffusion of antigenic materials from the eggs to the surrounding tissues was facilitated by a fenestrated shell; subsequent observations (discussed below) have provided abundant confirmation of this concept, which is the basis for the circumoval precipitin diagnostic test for schistosomiasis. The production and secretion of antigenic materials by schistosome eggs was aptly summarized by Warren (1980) and has been discussed by many authors subsequently.

A number of structures, including Reynolds' layer, von Lichtenberg's envelope, Lehman's Lacuna, and Cheever bodies, were described by Neill et al. (1988).

Composition

The Shell

Ziehl-Neelsen Staining. Following several early observations in Brazil, Lichtenberg and Lindenberg (1957) rediscovered that eggs of *S. mansoni* in sputum smears were stained by the Ziehl-Neelsen method for acid-fast bacteria. Shells stained strongly in tissue sections and in stool smears; contained miracidia stained only when the shell was intact. Lichtenberg and Lindenberg (1957) concluded that a specific alcohol-acid-fast substance exists in the shell of the *S. mansoni* egg. Eggs of various other trematodes and cestodes were subsequently tested by this method (review in Muller and Taylor 1972) and it was found that the eggs of some schistosomes (including *S. japonicum*) were positive and those of others negative. It was confirmed by Muller and Taylor that eggshells of *S. mansoni*, *S. rodhaini*, *S. mansoni-S. rodhaini* hybrids, and *S. intercalatum* were all deeply stained, whereas those of *S. haematobium*, *S. mattheei*, and *S. bovis* were not. The primary use of this method was thought to be differentiation of human infections of *S. intercalatum* from those of the other African terminal-spined schistosomes. As it had previously been determined that vitelline glands and undifferentiated eggshells of *S. haematobium* did stain by the Ziehl-Neelsen technique, Muller and Taylor considered that variations in the degree of acid-fastness of mature shells was due to the predominance of unsaturated fatty acids of high molecular weight in the positive-staining shells.

Phenolic Substances. Studying the distribution of phenolic substances and phenolase in *S. japonicum*, Ho and Yang (1973) subjected schistosomula and adult worms to a battery of chemical substrates. Only mature female worms oxidized p-cresol and tyramine rapidly, whereas o-catechol, L-tyrosine, and 3,4-dihydroxyphenylalanine were oxidized more slowly. Phenols and phenolase were found only in mature vitelline cells and in the shells of eggs in utero; the authors concluded that the eggshell of *S. japonicum* might arise from a quinone-tanned protein system.

Other Materials. Preparations of purified eggshells of *S. mansoni* and *S. japonicum* were made by Byram and Senft (1979). The most abundant amino acid was glycine, accounting for 37% of *S. mansoni* and 45% of *S. japonicum* amino acids; aspartic acid, lysine, and serine were also relatively abundant. Carbohydrates accounted for 7.5% to 10% of the shell weight in both species, with glucosamine as the principal amino sugar.

The Embryo

Lipids. Eggs of *S. mansoni* isolated enzymatically from tissues of infected mice were analyzed for lipids and other materials after sonic disintegration (Toro-Goyco and Rosas del Valle 1970). A lipid-rich lipoprotein fraction yielded a thin-layer chromatographic pattern similar to that shown by the lipids of whole eggs. Esterified cholesterol, triglycerides, fatty acids and their esters, and cholesterol were primary components, and phospholipids were found only in minute amounts. The most abundant fatty acids were palmitic, stearic, oleic, and linoleic, accounting for 90% of the total fatty acids. Linolenic and arachidonic acids constituted a further 5%. The lipids of *S. japonicum* eggs were studied subsequently, with different findings, by another group of workers (Brown et al. 1977, Smith et al. 1977) who found phospholipids (33.3%) and cholesterol (free sterol) (21.5%), free fatty acids (18.6%), triglyceride (16.6%), and sterol esters (10.1%) to be the major classes. Gas-liquid chromatography of fatty acid methyl esters showed fatty acids varying in chain length from 12 to 26 carbons, and with no to 6 double bonds; the saturated:unsaturated content was 40%:50%. The main phospholipid classes were phosphatidyl choline and phosphatidyl ethanolamine, although trace amounts of sphingomyelin and lysolecithin were also found. The high phospholipid-low sterol ester content reported from *S. japonicum* eggs (Smith et al. 1977) is in direct contrast to the opposite ratio reported earlier from *S. mansoni* eggs (Toro-Goyco and Rosas del Valle 1970). In another study on *S. mansoni* egg lipids (Smith et al. 1971) phospholipids and polar lipids constituted 21% of total lipids. Phosphatidylcholine, phosphatidylethanolamine, and lysophosphatidylethanolamine were major fractions, with lysolecithin, phosphatidylserine, and cardiolipin also present. A lysophosphatide mixture, adsorbed onto 60 μm particles of bentonite, elicited inflammatory lung granulomata when injected into sensitized or unsensitized mice.

Amino Acids and Proteins: Soluble Egg Antigen (SEA). Preliminary analysis on sonicates of *S. mansoni* eggs showed a "considerable amount" of free amino acids, of which 6.2% were identified as aromatic and 34.4% as basic (Toro-Goyco and Rosas del Valle 1970). Incubation of eggs of this species in 1.7% saline resulted in release of large amounts of proline, thought to be associated with stimulation of fibrosis in schistosomal granulomas (Isseroff et al. 1983); substantial amounts of alanine, arginine, and other amino acids were also released.

Extracts of the eggs of *S. mansoni* were found by Asch and Dresden (1979) to have acidic thiol proteolytic activity similar to that of cathepsin B. It was speculated that these proteinases may be involved in (1) penetration of eggs through hose intestinal tissues, (2) nutrition of the developing miracidial embryo or intramolluscan sporocysts, (3) escape of the miracidium from the shell, or (4) penetration of the miracidium into the snail. Immunological consequences of enzyme leakage were considered.

The schistosome egg contains a complex of protein materials at least some of which can be released to the tissues of the host. More than 50 years ago Hoeppli (1932) stated that "It has long been known that eggs of *Schistosoma japonicum* embedded in the tissues show some secretion . . ." and surmised that such secretion originates from the miracidium and passes through the intact eggshell. Subsequent authors have often referred to this flow of secretion as the "Hoeppli phenomenon." Working in Brazil with *S. mansoni*, Kloetzel (1967a, b, 1968) showed that a collagenaselike proteolytic enzyme was released from purified eggs isolated from mouse intestines. This enzyme was active on both gelatin and azocollagen substrates. Additional enzymes of schistosome egg origin have been described by Smith (1974) and by Asch and Dresden (1979). An immunologic study of *S. mansoni* eggs (Capron and Vernes 1968) identified at least 20 fractions, of which 13 were shared with adults and 7 specific to the egg stage. Shortly thereafter Boros and Warren (1970) and Hang et al. (1974) demonstrated that soluble egg antigens (SEA) contribute to granuloma formation by sensitizing murine hosts. Several studies (e.g., by Boctor et al. 1979; Warren 1980) have described numerous soluble egg antigen secretions, some which play a role in the pathogenesis of schistosomiasis and the elicitation of massive inflammatory responses to the eggs. These substances include proteins, polysaccharides, and glycoproteins, and they may vary considerably in concentration among schistosome eggs of different geographic origin (Hamburger et al. 1982). Monoclonal antibodies raised against certain components of the egg of *S. mansoni* have been found to protect mice against cercarial challenge (Harn et al. 1984). Further consideration of these substances is beyond the scope of the present review.

4

Sexual and Conjugal Biology

The conjugal relations between male and female schistosomes and the natural history of their sexual development represents one of the most interesting chapters in all of parasitology. Many studies have aimed at understanding how this system functions within the constraints and opportunities presented by infection of vertebrate hosts with naturally occurring schistosome cercariae.

Sex Ratio

Early studies on naturally infected snails were done in Japan, where Tanabe (1919) found that 26 of 31 animals exposed to cercariae from individual naturally infected *Oncomelania* snails had worms all of one sex. Also in Japan, Sugiura (1934) found *Oncomelania nosophora* naturally infected with 47.6% male, 38.2% female, and 14.2% mixed cercariae. Severinghaus (1928), in China, collected thousands of local *Oncomelania,* which yielded a total of 12 male, 12 female, and 3 bisexual infections. Some years before, Faust (1927) had had ten thousand *O. hupensis* collected for him in the vicinity of Soochow over a period of 7 months, during a prolonged dry period. Of these only 20 were found infected. Rabbits and dogs exposed to the cercariae had only male worms at necropsy, although previous collections from the same area during more seasonable weather had yielded worms of both sexes.

The prosobranch molluscan hosts of the Asian schistosomes are themselves dioecious, unlike the hermaphroditic *Biomphalaria* and *Bulinus* hosts of the African and American species of schistosomes infecting humans. It may be, therefore, that the sex of the emergent cercariae is influenced by the sex of the snail host. This possibility was investigated by El Raggal (1959), who detected no such effect in laboratory reared *O. hupensis* "unimiracidially" infected with *S. japonicum*, both originating in Zhejiang province: 32 male snails had 16 male, 12 female, and 4 bisexual infections, while 19 female snails had 10 male, 7 female, and 2 bisexual. The most elaborate statistical analysis of cercarial sex in relation to miracidial exposure and sex of snail host was made by Tanaka and Matsuda (1973). These authors set up 8 experiments with about 70 *O. nosophora* snails in each, exposed to 1, 2, 3, 5, or 10 miracidia. The theoretical probabilities of unisexual or bisexual infections were estimated with the assumption of equal infection and development rates for either parasite sex in male or female snails. The observed ratios were not significantly different from those predicted by chance in 6 of the 8 experiments, and the occurrence of bisexual infection was not less than the theoretical probability. The average probability of successful development of each miracidium overall was 0.22; the proportion of males was 0.37 to 0.64, with an average of 0.48, and was independent of the sex of the snail host.

Various researchers have also studied the sex of cercariae shed from individual field-collected snails naturally infected with *S. mansoni*. Thus 93 guinea pigs were exposed percutaneously to cercariae from individual naturally infected snails collected in Puerto Rico in an area with "constant and intense pollution of the water" (Maldonado and Herrera 1950). Resulting infections were male in 19, female in 10, and bisexual in 40 animals. At about the same time, in Belo Horizonte, Brazil, Paraense and Malheiros Santos (1949) exposed rabbits, guinea pigs, rats, and mice individually to cercariae shed from 53 naturally infected local *B. glabrata* snails. The snails had also been collected in an area of intense pollution, having a 29% overall infection rate, which is very high for a natural population. Resulting infections were: 17 male, 14 female, 6 both, and the remainder none or indeterminate. These authors also infected animals three times at weekly intervals with cercariae from three individual field-infected snails. One snail produced a bisexual infection the first week and only females the second; another produced bisexual infections 3 times, and the third, males twice and a bisexual infection the third time. Paraense (1949) then examined 400 "*Australorbis olivaceus*" collected on the grounds of the Instituto Oswaldo Cruz in Rio de Janeiro, finding a 8.25% infection rate. Two dozen guinea pigs were each infected with cercariae from a separate snail, all resulting infections being unisexual (13 males, 11 females).

The effect of the seasons on infection in *Biomphalaria* was studied in Venezuela by Pifano (1948), who found that 10 snails collected during the rainy season gave rise to 205 males and 187 females, but that 18 snails collected during the dry season yielded 590 males and no females. This unusual sex ratio, recalling the earlier report of Faust (1927), stimulated an experiment in which 10 snails were collected during the rainy season and 5 mice infected with the pooled cercariae, resulting in 174 males and 147 females. The same snails were then exposed to sunlight over 10 days to produce a 50% reduction in container water volume. Four mice exposed to cercariae from 10 of those snails were found to have 148 male and 10 female *S. man-*

soni. In contrast, Jaffe et al. (1945), also working in Venezuela, reported that the percentage of male and female infections in local snails was about the same regardless of season. It remains to be determined whether the report of Pifano represents a statistical peculiarity or whether female sporocysts are more susceptible to the effects of dryness.

Few investigators have studied the total numbers of male and female cercariae shed by snails. Rowntree and James (1977), working with unimiracidial infections of *S. mansoni,* found that in their hands a) more snails became infected with female parasites, b) male-infected snails tended to die earlier, c) male- and female-infected snails showed no difference in number of cercariae shed, and d) cercariae of both sexes were shed after the same interval. For *S. intercalatum,* by contrast, Frandsen (1977) reported that more male than female cercariae were produced because male-shedding snails lived and shed longer. *Bulinus globosus* infected with a Zaire isolate shed for 119 days, producing 15,500 cercariae (male), versus 32 days and 3,300 cercariae (female). Similarly, *B. forksalii* infected with a Cameroon isolate shed for 94 days (male) and 62 days (female) with corresponding numbers of cercariae produced. Wright and Knowles (1972) found that in *S. haematobium* of Mauritian and Iranian origin male cercariae matured faster in *Bulinus* snails than did females. These authors also confirmed an interesting observation by Paperna (1968) that two isolates of *S. haematobium* from Ghana that normally developed in *Bulinus rohlfsi* and *B. globosus* could each infect the other species, but with a tendency in the "wrong" host to produce all male cercariae. This suggests a complex genetic interaction between parasite and snail that may influence the nature, including the sex, of resultant cercariae.

The frequent reporting of male schistosome preponderance stimulated Liberatos (1987) to carry out the most elaborate study of sex ratio in *S. mansoni.* He postulated the following possible mechanisms for persistently skewed sex ratios:

1) other than a 50:50 sex ratio of miracidia; 2) differential mortality between male and female miracidia; 3) differential infectivity between male and female miracidia; 4) differences in lengths of prepatent periods of male and female infections in host snails; 5) differential longevity of male and female snail infections, affecting the numbers of cercariae produced; 6) differential production of numbers of cercariae by male and female sporocysts; 7) differences in sizes of snails with male and female parasites, affecting the numbers of cercariae produced; 8) differential mortality between male and female cercariae; 9) differential infectivity between male and female cercariae. (Liberatos 1987)

In testing each of these possibilities, he found that the sex ratio of miracidia was 50:50 and their longevity was the same (about 10 hr). However, male miracidia were more successful in snails, developing to cercarial shedding in unimiracidial infections in a ratio of 57 males:34 female. Prepatent periods, snail size and longevity, and the total number of cercariae produced were not significantly different. The average number of cercariae shed was about 360 per day, or about 32,000 over the lifetime of an infected snail.

Under the experimental conditions employed male cercariae lived a mean of 21.3 hours and females 25.0 hours, not significantly different; however, differences in longevity of pools of cercariae shed by individual snails were significant at the .001 level, independent of sex, presumably a genetic character; compare with the observations of Cohen and Eveland (1984) (following paragraph). In mice given carefully counted numbers of male and female cercariae, about 16% of male and 9% of female parasites reached adulthood, a relatively low proportion.

Using the technique of sporocyst transplantation from infected to uninfected *B. glabrata*, Cohen and Eveland (1984) were successful in transferring 5 male-producing and 5 female-producing clones of *S. mansoni* over 3 to 6 passages each for up to 2 years. Beginning 5 weeks from the onset of cercarial emergence, these authors found an.average shed of 3,900 cercariae/snail from male clones and 1,300 cercariae/snail from female clones at each shedding period ($p<.001$ by t-test). The cercarial shed varied from clone to clone, but was relatively stable within each clone, suggesting that the number of cercariae released at each shedding period is under genetic control.

As unexpected, presumably genetic variation in snail infectivity in *S. mansoni* was reported by Lagrange (1954), who studied two strains of *S. mansoni* over many years. In one strain he found that individual snails were unable to shed cercariae of both sexes, always producing unisexual infections in mice, while another could produce bisexual infections in mice exposed to cercariae from indvidual snails.

A careful study of the numbers of male and female *S. mattheei* developing from known numbers of cercariae of each sex was made by van Rensburg and van Wyk (1981), who exposed 12 sheep each to 100 male and 100 female cercariae, 96.6% of which penetrated into the animals. Of the 1,200 cercariae of each sex used, 609 male and 800 female worms were recovered. The authors concluded that the usual majority of male *S. mattheei* in sheep is due to exposure to a larger number of male cercariae, not to their greater infectivity.

Identification of Sex

The development of a simple and reliable technique to distinguish the gender of cercariae is of critical importance for experiments in which the sex of infecting schistosomes must be known. Direct determination of parasite gender eliminates the necessity for tedious test infections, in which mice are exposed to cercariae shed by individual unimiracidially infected snails, and after 1 month are killed to determine the sex of the schistosomes carried by that snail. In the interval the snail may die and the work becomes useless. The method of Liberatos and Short (1983), for cytological differentiation of male and female cercariae of *S. mansoni*, is based on the presence or absence of W-chromosome constituitive heterochromatin in cercarial tail interphase nuclei. The technique was employed in my own laboratory over a period of several years, during which hundreds of unimiracidial infections were produced in *Biomphalaria glabrata* snails. A skilled technician (my wife, Natalicia), always using freshly made reagents and following the protocol with scrupulous care, correctly determined the sex of cercariae shed by these snails in every single instance, as demonstrated by 100% recovery of the proper sex in subsequent animal infections. The only tools necessary were a balance to weigh reagents, some beakers and bottles, a warming bath and thermometer, slides, and a microscope with oil immersion lens. By contrast, Spotila et al. (1987), Webster et al. (1989), and Walker et al. (1989a) have described DNA probes intended to accomplish the same goal but requiring a laboratory full of gel electrophoresis apparatuses, restriction enzymes, plasmid vectors, nitrocellulose filters, radioisotopes, and the full paraphernalia of modern molecular biology, using infinitely more expensive, complex, and prolonged procedures. The latter two groups had perhaps independently found the same 476 base-pair fragment which Walker et al. (1989a) ascribed to "the satellite DNA in the heterochromatin region of the W. chromosome"—which appears to be the precise area that is visualized simply, rapidly, and directly, using the method of Liberatos and Short (1983).

Intramolluscan sporocysts in unimiracidially infected snails can be sexed by the method of Grossman et al. (1980). It can be demonstrated that in females only, one homologue of the first pair of chromosomes has noncentromeric heterochromatin found in the mitotic metaphase stained with the standard Giemsa technique.

Pairing and Copulation

Most infections with pooled cercariae from several snails will yield bisexual infections in the vertebrate hosts. Observations on the sex ratio of adult schistosomes from naturally and experimentally infected animals are summarized on Table 4–1.

Where both sexes of schistosomes are present, pairing regularly ensues even when the male and female are of different species. This alleged "instinct to pair" is exhibited by S. japonicum 10 to 12 days before attainment of sexual maturity (Faust and Meleney 1924).

When Michaels (1969) allowed transected halves of S. mansoni to pair in vitro, like halves almost always paired normally and unlike halves almost always paired in an inverted sense (>90%). The anterior halves of males did not pair readily with intact females, but posterior halves did so. It was postulated that worms possess linear receptors used to determine mating position. Platt et al. (1991) reported that in Griphobilharzia amoena from Australian crocodiles, females in the gynecophoral chamber were always found in a reverse position to the males. The normal mating position of most known schistosomes is parallel, anterior/anterior. In S. japonicum, Faust and Meleney (1924) reported finding rare pairs in an inverted position in vivo. In S. mansoni, 12% of females were found by Faust et al. (1934) in reverse position, but in all cases their heads were "directed towards the terminus of the venules."

Michaels (1969) considered pairing to be an active process by both part-

TABLE 4–1 Sex Ratio of Adult Schistosomes

Species	Host	Male:Female n:1	Natural/ Experimental	Comment	Reference Note
S. mansoni	Mouse	2.04	Exp	Castrated host	1
	Mouse	2.65	Exp	Intact host	2
	Rabbit	varied	Exp	1.23 to 11.3	
	Monkey	varied	Exp	1.41 to 1.63	
	Rattus	1.08	Nat	Guadeloupe	3
	Mouse?	4.03	Exp		4
	Mouse	1.5	Exp		5
	Hamster	1.22	Exp	M 8% to 10% more	6
	Mouse	Excess of M	Exp	Exposed 1:1 cerc.	7
	Guinea Pig	1.63	Exp	50M or 50F/ mouse given separately	8
	Mouse	143M:79F	Exp		
S. haematobium	Hamster	Excess of M		Common ratio	9
	Various	5	Exp		
	Hamster	1.56	Exp	Iran	10
	Hamster	2.23	Exp	Nigeria	11
	Hamster	2.85	Exp	Durban	12
	Hamster	1	Exp	Egypt	13
	Hamster	8	Exp	Tanzania	
S. intercalatum	Mouse	1.6	Exp	Cameroon	14
	Mouse	1.8	Exp	Zaire	15
	Mouse	1.6, 2.2	Exp		16
	Male Hamster	2.5			
	Female Hamster	3.63			

Species	Host	Type	Value	Notes	Ref
S. bovis	Mouse	Exp	2.57		17
	Guinea Pig	Exp		737:1 and 295:2	18
	Male Guinea Pig	Exp	1.71		
	Female Guinea Pig	Exp	2.94		
S. mattheei	Mouse	Exp	0.76	No pref	19
	Sheep	Exp		Exposed 1:1 cerc.	20
	Male Hamster	Exp	0.67		21
	Female Hamster	Exp	0.78		
S. margarebowiei	Male Hamster	Exp	1.40		22
	Female Hamster	Exp	1.06		
S. nasalis	Guinea Pig	Exp	all Male		23
S. spindale	Guinea Pig	Exp	all Male		24
	Guinea Pig	Exp			25
S. japonicum	Cattle	Nat	1.33	Majority male	26
	Mouse	Exp	1.13		27
	Guinea Pig	Exp	1		
Schistosomatium douthitti	*Peromyscus*	Exp	1.82	24 mice	28
Orientobilharzia turkestanicu	Various	Exp	0.6 to 3.9	See original	29
Ornithoobilharzia canaliculatum	Birds	Nat		Majority male	30
Dendritobilharzia pulverulenta	Birds	Nat	100:72		31

Reference notes: 1, Berg 1953; 2, Faust et al. 1934; 3, Imbert-Establet 1982; 4, Paraense and Malheiros Santos 1949; 5, Standen 1949; 6, Standen 1953; 7, Stirewalt et al. 1951; 8, Liberatos 1987; 9, Purnell 1966; 10, Rowntree and James 1977; 11, Webbe and James 1972; 12, Wright and Bennett 1967a; 13, James and Webbe 1973; 14, Bjorneboe and Frandsen 1979; 15, Frandsen 1977; 16, Wright et al. 1972; 17, Brumpt 1930a; 18, Brumpt 1936; 19, Lengy 1962; 20, van Rensburg and van Wyk 1981; 21, Wright et al. 1972; 22, Southgate and Knowles 1977; 23, Rao 1938; 24, Rao 1938; 25, Fairley et al. 1930; 26, Fujinami 1907; 27, Vogel and Minning 1947; 28, Short 1952a; 29, Massoud 1973b; 30, Witenberg and Lengy 1967; 31, Ulmer and vandeVusse 1970.

ners. However, some male schistosomes have a propensity to pair with or clasp almost any object the approximate size and shape of a female schistosome. In worms maintained in culture media I have sometimes observed males clasping cotton fibers. Purpose-made linear fragments of alginate materials in culture medium were held avidly by both unisexual and formerly paired male *S. mansoni* that had been removed from mouse hosts (Basch and Nicolas 1989). These observations suggest that clasping is a thigmotactic behavior independent of pheromonal signals from females. Striebel (1969) followed pairs of *S. mansoni* or *S. japonicum* in mice treated with niridazole. It was found that females died and became necrotic but males were fully motile and remained in copula with the dead females.

Many authors have found that re-pairing of separated sexes occurs readily in vitro. For example, Michaels and Prata (1968) found that of 140 tubes containing single male and female *S. mansoni* perfused from mice (not original partners), 20.7% had paired by day 1, 84.2% by day 2, 94.5% by day 3, and 100% by day 5.

Pairing of one male and one female appears to be the norm in schistosomes, but in cultures of *S. mansoni* it is common to find single female worms held by two males, one at either end. In an experimental mixed infection, Armstrong (1965) found 29 females of *S. mansoni* paired simultaneously with one large male *Heterobilharzia americana;* all the females, as well as the male, appeared sexually functional. A similar phenomenon may occur within a single species: Chu and Cutress (1954) reported that in *Austrobilharzia variglandis* polygamy was the rule, with 3 to 8 females occupying the gynecophoral canal of most males. In these cases, some of the females were small and retarded. I have found that the clasping of a single female by two males in mouse infections is more common in *S. japonicum* than in *S. mansoni.*

To demonstrate that copulatory pairing is not universal in schistosomes, Ulmer and vandeVusse (1970) stated that although *Dendritobilharzia pulverulenta* had never been found in copula, all mature females from natural infections were found to be gravid.

The biochemistry, physiology, and immunology of pairing has been relatively little studied. It was found by Aronstein and Strand (1984) that 11 of 32 antigenic glycoproteins of *S. mansoni* were recognized only by the antiserum from bisexually infected mice, and not by antisera from mice infected with either sex alone. It was concluded that certain pair-dependent antigens were produced, which may be relevant to the immune mechanisms involved in pathogenesis and resistance in mice.

Interactions of Male and Female Worms

The dioecious nature of schistosomes, unique within the trematodes, has enabled this group to adopt a variety of conjugal relationships, the best known of which is the inability of some females to grow and mature in the absence of continued physical contact with an appropriate male. It must be emphasized that such female dependence upon the male is not absolute even within the genus *Schistosoma,* and is not exhibited at all in some genera such as *Schistosomatium.* More significantly, this relationship is virtually unexplored in the vast majority of species within the family, for example, those of birds, leaving abundant scope for further investigation.

The diversity and evolution of reproductive strategies in schistosomes has been discussed by Basch (1990), who outlined several functions of the male (Basch 1990). Clearly, insemination and fertilization in the classical mode serves to restore diploidy and contributes to the diversity of the gene pool, facilitating adaptation and evolutionary change. However, as will be discussed, insemination may be unecessary either for female growth and development or for production of viable miracidia. In those schistosomes in which the musculature of the immature female is poorly developed, the male is necessary as a tractor, to transport the weaker female to sites of oviposition.

Early elucidation of male-female relationships arose primarily from the work of a cohort of Japanese investigators studying *S. japonicum* (citations and brief review in Moore et al. 1954). Severinghaus (1928), an American working in China, was one of the first workers to note that female worms failed to grow and mature when the male was not present, concluding that "it is conceivable that hormones produced by the male may in some way be connected with the phenomenon." He found also that males in unisexual infections grew more slowly and remained smaller than paired males in bisexual infections. Subsequently it was seen by Li (1958) that excess unpaired males of *S. japonicum* from bisexual infections were intermediate in size between unisexual and paired males. Unpaired males of certain other species, such as *S. bovis* (Lengy 1962b), do not attain full size, whereas those of *S. mansoni* appear to develop normally, irrespective of the presence of female worms. In these species, as well as in *S. haematobium* (Brumpt 1936), unisexual females are severely stunted.

In *Schistosomatium douthitti* (Fig. 4–1) Short (1952a) showed that males are needed only for the females to attain full size, not for induction of their maturation.

The studies of Vogel (1941a) established that female *S. mansoni* are not stimulated to mature until the males have themselves become mature. This was done through series of experiments in which mice were infected separately with male and female cercariae at various intervals. Vogel (1941a) divided female development into two phases: first, until 1 month of age (development of genital anlage, when the female was less than half the mature body length) the female was considered more or less independent of the male. Second, from the beginning of the second to the middle of the third month (sexual maturation, full growth) the female must be carried within the gynecophoral canal of the male. The necessity for pairing occurred as soon as 19 days following superinfection with male cercariae (Standen 1953b).

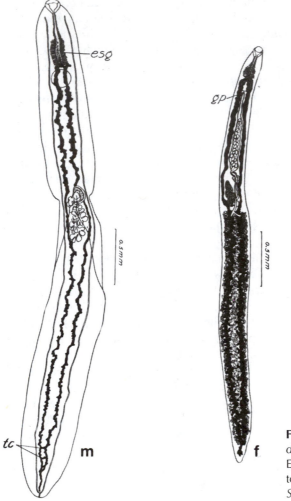

Figure 4–1 *Schistosomatium doüthitti,* adult male and female. Esg, esophagus; gp, genital pore; tc, transverse commissure of gut. *Source:* Price 1931.

Moore et al. (1954) found that the rate of growth and maturation of females was not altered by introduction of males into a mouse with a long-standing female infection, or vice versa, as compared with simultaneous infection of cercariae of both sexes.

The genetic makeup of the male worms modulates the rate of development of the pair. Hsü et al. (1962) found that in *S. japonicum*, Chinese and Taiwanese isolates had longer prepatent periods than did Japanese and Philippine isolates. Chinese or Taiwanese (slow) females were found to develop more rapidly when crossed with Japanese or Philippine (fast) males. It was concluded that the maturation of the females was accelerated by the early maturation of the males, which thereby controlled the rate of female development.

Interchange of Materials

Copulatory: Insemination. To determine the significance of insemination for female maturation, Shaw (1977) placed ex-vivo unisexual female *S. mansoni* into culture medium in vitro and added ex-vivo mature males. Most females were paired after 24 hours, and within 5 to 6 days these displayed a variable amount of vitelline material visualized with fast red B and transmission electron microscopy. Some even had a poor egg in the ootype, complete with shell showing fenestrations and microspines. About 85% of these females had not been inseminated. It was concluded that insemination per se is not a "major trigger" for development.

Michaels (1969) conducted experiments in vitro in which males were made anorchid by prolonged maintenance in vitro or by exposure to X-rays, and then paired with freshly isolated female *S. mansoni*. It was found that such males with deteriorated testes stimulated the rate of egg-laying and development as well as did intact freshly isolated males. Thus neither mating nor stimulation of the rate of oviposition depended upon intact testes, viable sperm, or something secreted with the sperm.

To determine the frequency of copulation, Nollen et al. (1976) incubated males of *S. mansoni, S. japonicum,* and *S. haematobium* in medium containing tritiated thymidine and implanted labeled males and unlabeled females into mice. By day 6 labeled spermatozoa were seen in the seminal vesicles of all species. On the same day, if the male was paired, the female often contained labeled sperm in the seminal receptacle, indicating that insemination occurs often.

Nongametic. The interchange of soluble materials between copulatory partners has been suggested by numerous authors for more than half a century, but despite several specifically-directed investigations the situation remains obstinately ambiguous.

Direct transtegumental absorption of low molecular weight compounds was proven by Asch and Read (1975a) who prevented ingestion by ligating male *S. mansoni* across the body just posterior to the oral sucker. The worm bodies were then suspended into medium containing radiolabeled glycine or proline, which were taken up across the tegument. In a more elaborate study Asch and Read (1975b) showed that in adult male *S. mansoni* cysteine is taken up by diffusion alone, proline by active transport alone, and glycine, methionine, arginine, glutamate, and tryptophan are absorbed by diffusion as well as by at least five distinct transport systems. The passage of materials through the schistosome tegument is therefore a very complex process and, if minute amounts of materials might be transferred from worm to worm in an instantaneous process, the present state of confusion can be readily understood.

Noting far heavier uptake of radiolabeled proline by male *S. mansoni*, Senft (1968) and Grant and Senft (1971) speculated that the gynecophoral canal may serve in transfer of this amino acid from male to female worms. On the other hand, the observation by Chappell (1974) of only minor differences between the sexes in uptake of methionine "would seem to contrast with the suggestion of Senft (1968) that the male may furnish the female with amino acids in vivo."

It was suggested by Atkinson and Atkinson (1980b) that the male *S. mansoni* retains little of the protein that it produces in greatest abundance, and that this material, of approximately 66,000 molecular weight, was found in but not synthesized by the female. This conclusion was based on incubation of separated males in medium containing radiolabeled leucine, followed by re-pairing with an unlabeled female, reseparation, and homogenization of each partner. Polyacrylamide gels of the homogenates were then autoradiographed to locate the label. A similar protocol was followed by Popiel and Basch (1984b), who also undertook autoradiographs of sectioned worms in an attempt to localize label transfer anatomically, but no significant transfer of any polypeptide was detected. This seeming contradiction in results remains to be resolved.

Exchange of Carbohydrates. Carbohydrate metabolism of *S. mansoni* is exceedingly vigorous. As long ago as 1950 Bueding and Koletsky found that these worms utilize about one-fifth of their weight in glucose every hour. Rogers and Bueding (1975) reported that glucose was taken up through the tegument rather than orally in *S. mansoni,* and that this "is consistent with the hypothesis that in copula the male schistosome provides glucose to the female." Rogers and Bueding (1975) cited unpublished dissertations by Holden and Lennox that suggested transintegumental transfer of nutrients from male to female schistosomes. It had been found by Bueding and Koletsky (1950) that glycogen content of male *S. mansoni* tended to increase with age. In males roughly 14 to 30% of the dry weight consisted of glycogen; in females the amount was 2.7 to 5%. Cornford and Huot (1981) found that the rate of glucose assimilation by male and female *S. mansoni* was significantly greater in copulating than in separated flukes. Also studying the comparative utilization of glucose in separated and mated schistosomes, Cornford and Fitzpatrick (1987) reported that 2-deoxyglucose (2-DG) is taken up more rapidly, but phosphorylated much more slowly, than glucose by both *S. mansoni* and *S. japonicum.* In both species, mated males in vitro phosphorylated 2-DG at a greater rate than unmated (recently separated) males, and mated females consumed and phosphorylated more glucose than did recently separated females. All rates were far higher in *S. japonicum* than in *S. mansoni.* In *S. mansoni, S. japonicum,* and *S. haematobium* glycogen content was greater in unpaired than in copulating males, suggesting to Cornford and Huot (1981) that during pairing the female depletes stored glycogen from the male. These authors also demonstrated transfer of labeled glucose from male to female and suggested that a considerable portion of the female's glucose supply comes from her male partner. Cornford and Fitzpatrick (1985) demonstrated that male to female transfer of glucose is not inhibited by ouabain.

They suggested that the transfer follows a concentration gradient between males and females, and is therefore not an active transport process.

Cholesterol and Lipids. The independent observations of Haseeb et al. (1985) and of Popiel and Basch (1986), both show reciprocal transfer of cholesterol between male and female *S. mansoni*. Male worms in RPMI-1640 culture medium were reported by Silveira et al. (1986) to incorporate ^{14}C-labeled cholesterol and transfer it and its metabolites to females; females could take up the labeled cholesterol and transfer it, but not its conversion products, to males. Immature schistosomula could take up, but not convert, labeled cholesterol.

Other investigators have suggested that factors released by females can affect the physiology of males. For example, Haseeb et al. (1989) concluded, on the basis of in vitro experiments, that males from unisexual infections showed increased lipid accumulation and lipase activity when incubated for an hour with formerly paired mature females (see also the discussion on lipids, below).

Amino Acids and Peptides. The tripeptide glutathione, [L-gamma-glutamyl-L-cysteinylglycine], which comprises about 95% of nonprotein thiols in schistosomes, occurs in 1.7-fold greater concentration in males from single-sex infections than in paired males, and biosynthesis in vitro is 5 times higher in unpaired than in paired males (Siegel and Tracy 1988, 1989). These authors suggested that glutathione biosynthesis in males may be regulated by a response stimulated by the physical presence of the female in the gynecophoral canal.

The apparent permeability of the adult worm tegument may differ in the paired and separated state. Cornford (1985) reported for *S. mansoni* that neither males nor females took up acidic amino acids (aspartate and glutamate) when paired, but that both did so when incubated in isolation. In *S. japonicum* neither paired nor separated worms took up glutamate, but in *S. haematobium* both mated and unmated adults showed saturable uptake of aspartate. The reproductive significance of these findings remains uncertain.

Male Stimulation of the Female

The role of the male worm in inducing female development is certainly not the same for all species of schistosomes. In the forms most studied, *S. mansoni* and *S. japonicum,* the manner in which the male exerts its influence is not completely understood. Among the possible mechanisms, the provision of hormonal or nutritive material by the male to the female has been considered many times (Severinghaus 1928, Moore et al. 1954, Vogel 1941b). The role of lipids in particular has been suggested. It is known (Rumjanek and Simpson 1980) that live adult *S. mansoni* maintained in vitro can absorb and incorporate into membranes radiolabeled arachidonic acid, linolenic acid, phosphatidylcholine, tripalmitylglycerol, and cholesterol from the culture medium. Arachidonic acid was incorporated most readily in this series, but no evidence of prostaglandin synthesis was found. Perhaps the most persuasive case for the reproductive significance of lipids was made by Erasmus and Shaw (1977) and Shaw et al. (1977), who reported that acetone or ether extracts of male S. *mansoni* stimulated vitelline cell development in ex-vivo unisexual females maintained in vitro. Female development was assessed by fast red B staining and ultrastructural study. This work was confirmed by Popiel and Erasmus (1981a) who found that the rate of tyrosine uptake by unisexual females in culture was increased by addition of male worms or male worm lipid extract.

Den Hollander and Erasmus (1985) placed immature unisexual females of S. *mansoni* in vitro with mature males. Within a day after pairing these females took up approximately twice the labeled thymidine of isolated females, indicating augmented synthesis of DNA. Those females that were incubated with males but did not pair increased their relative incorporation of labeled thymidine by 35 to 45% in comparison with isolated females. Basch (1988) perfused unisexual female S. *mansoni* from mice and paired them for one week in vitro with 1) intact males; 2) segments of males cut transversely into thirds; or 3) no males. All were incubated with radiolabeled tyrosine or thymidine. The portions in contact with cut male segments showed significantly higher uptake of thymidine than the noncontact halves, indicating increased DNA synthesis and cell division, but dot-blot hybridizations with a female specific single-stranded cDNA failed to detect production of the corresponding mRNA in females paired with male segments. Noncontact halves showed greater uptake of tyrosine.

The avid uptake of cholesterol from host blood (Smith et al. 1970) suggests that steroid hormones may be significant in schistosomes, but there is little evidence for this. Studies by Nirde et al. (1983, 1984), Torpier et al. (1982), and Basch (1986a) have shown by chemical and immunocytochemical means the presence of ecdysone-like compounds, but their role in schistosome biology remains speculative.

It was suggested by Gupta and Basch (1987a) that unisexual females of S. *mansoni* are stunted at least in part because of malnutrition. Their poorly developed pharyngeal and gut musculature is considered inadequate to pump sufficient blood into their intestinal ceca. Therefore, one function of the male may be physical massage of the female to induce a sort of external peristalsis that results in filling of the female intestine with host blood, with subsequent digestion and growth. On the other hand, females of *Schistosomatium douthitti,* which can develop to maturity independent of male worms, must be capable of feeding on their own. Indirect evidence for this view included relative measurements of the uptake of radioactively labeled erythrocytes, the thickness of pharyngeal and intestinal musculature of unisexual and paired females of both species, and immunocytochemical studies of the secretion of acidic thiol proteinase digestive enzyme.

Other possible mechanisms of male stimulation are: transport of females to a site in the mesenteric veins where the microenvironment (such as host hormones) induces them to feed and grow; and stimulation of nerve endings in the female, which directly or indirectly causes the release of maturational neurohormones that act upon the reproductive system. No evidence exists in favor of these speculations.

Specificity of Reproductive Stimulation

Reproductive stimulation by males is not species specific. Interspecific "crosses" among *S. mansoni, S. japonicum,* and *S. haematobium* in all combinations were made a half century ago (Vogel 1941b), showing that females developed to maturity and even produced some eggs, always of the maternal type. Taylor (1970) carried out similar studies with several species of African schistosomes. Short (1948b) and Armstrong (1965) "mated" female *S. mansoni* with males of *Schistosomatium douthitti* by infecting mice with both, and found development of some eggs by the females. We (Basch and Basch 1984) have followed up on this work, and demonstrated that the female *S. mansoni* in such combinations are stimulated to mature, but not to grow, by pairing with *S. douthitti.* Further details are to be seen in the section on parthenogenesis on page 154.

In an attempt to determine whether the male to female stimulus is localized or propagated, Popiel and Basch (1984b) paired pieces of males of *S. mansoni,* transected with a scalpel at various levels, with intact unisexual females of the same species both in vitro and by implantation into hamsters. They found that only the region of the female in contact with the male segment developed some vitelline material. It was concluded that worm pairing, male stimulation, and the female developmental response are independent of central nervous control, that males have no centralized location for production of the female-stimulating factor, that vitelline gland differentiation requires local stimulation through male contact, and that this is not propagated throughout the worm.

Artificial "Females" to Capture Secreted Molecules. "Surrogate" females were created out of strands of flexible, inert alginate material of simple chemical makeup in order to absorb the putative stimulatory materials emitted by embracing males (Basch and Nicolas 1989). Adult male *S. mansoni* were permitted to clasp and "pair" with the strands, which they did fairly readily, producing hundreds of "pairs" that often remained for many days. The alginate strands were removed after several days, solubilized, and subjected to gel electrophoresis to separate the polypeptides that they had absorbed. If such materials could then be eluted from the separated fibers, important information might be obtained about the nature of the male-female relationship. Proteins were found in "paired" strands at approximately 40 and 46 kilodaltons that were never found in strands incubated in the culture medium alone. It is possible that these bands represent substances produced by males and transferred to females during pairing. What these substances are, and whether they produce any physiological effect in the females, remains to be determined.

Effects of Experimental Irradiation. The effect of irradiation on subsequent development and mating behavior of *S. mansoni* in mice was studied by Armstrong (1965). Cercariae irradiated at 1,200 rad yielded worms with widely variant damage, from death at the lung stage to apparent normalcy. Gross physical deformities were seen only in males. Irradiated females when paired with normal males developed to nearly full size but had a damaged reproductive system. Badly deformed and sterile irradiated dwarf males often paired with normal females. Such females became sexually mature but were stunted in size; normal females paired with full-sized but sterile males achieved "almost full" maturity. This work suggests that stimuli for growth and for sexual maturation are not the same.

Separation of Pairs

Natural; in vivo. There has been some disagreement about the permanence of pairing in vivo—whether the female schistosome leaves the male to deposit its eggs, or remains in permanent copula. Fairley (1920) believed that females of *S. mansoni* and *S. haematobium* in monkeys often left the male to deposit eggs, and Faust et al. (1934) stated that in *S. mansoni* permanent pairing appears to be the exception rather than the rule. The paucity of unpaired *S. mansoni* in mesenteric veins in bisexual infections has led Gönnert (1955a) and other authors to believe that pairing is permanent. Standen (1953a) stated that once the pair is in position near the intestinal wall it is extremely unlikely that the female ever leaves the male entirely, but merely reaches forward to deposit the eggs. In *S. japonicum* Faust and Meleney (1924) always found males and females in copula and reported that females were paired during oviposition. Li (1958) believed that pairing was "permanent in character" and the worms did not seem to change partners.

In mice experimentally infected with *Heterobilharzia americana*, mature flukes were found in copula after 72 days. Immature worms were concentrated near the celiac plexus, and mature pairs near the serosal surface of the intestine. After 120 days the flukes were concentrated in the portal vein near the liver and many females were no longer associated with males. The mouse is an unusual host for *H. americana*, which is found naturally in dogs, rabbits, raccoons, lynx, opossums, deer, and nutria from North Carolina to Texas (Goff and Ronald 1981).

Easy exchange of partners appears to be the rule in *Schistosomatium douthitti*. In one mouse Short (1952a) found 12 males and 31 females; all of the first 24 females examined had sperm in the seminal receptacle.

Experimental. Various methods have been suggested for separation of living schistosome pairs removed from animals. Bueding et al. (1967) proposed 2×10^{-4} M carbachol, Shirazian and Schiller (1982) placed the dishes containing worm pairs on ice, and Popiel and Basch (1984b) used dilute nembutal in medium.

Mature ovigerous paired female *S. mansoni* were removed from their hosts and implanted surgically into uninfected hamsters with or without their male partners (Clough 1981). Paired females continued to produce eggs normally, but in the unisexually transplanted females degenerative changes were noted in the vitelline gland within 3 days and in the ovary within 6 days. Even after 3 months of reproductive regression, females regenerated and produced normal eggs when reunited with mature males. Thus Clough (1981) demonstrated that separation of female *S. mansoni* from males leads to a reversible reproductive degeneration. The morphology and reproductive status of female *S. mansoni* after separation from males was also considered by Popiel et al. (1984b). More recently, Den Hollander and Erasmus (1985) found that cultured females of *S. mansoni* deprived of males showed a continual decrease in thymidine uptake, in comparison with still paired females.

Thioxanthone drugs such as lucanthone and hycanthone act far more severely upon male worms than upon females, particularly in certain geographic isolates of *S. mansoni*. Lee (1972) used 25 mg per kg of mouse body weight of hycanthone to kill 70% of male NIH-PR *S. mansoni*, whereas twice that concentration killed only 30% of females. Females surviving after treatment were stunted and pale but returned to normal proportions after a prolonged period of pairing with males from a post-treatment cercarial infection. However, these widowed and remarried females failed ever to produce eggs. Popiel and Erasmus (1982) found similarly that in *S. mansoni*-infected mice given oxamniquine, male worms were differentially killed. Remaining females began to show regressive changes from 8 days post-treatment, deteriorating to a state near that of unisexual females. Administration of additional oxamniquine had no further obvious effect on these females, suggesting that the regression was a result of loss of the males rather than a direct effect of the drug. As with Lee's work, it was found subsequently (Popiel et al. 1984b) that females stunted after exposure to a modest dose (25 mg/host kg) of oxamniquine were able to recover fully when reunited with normal male worms in vivo. Females from animals given 50 mg/kg of oxamniquine remained unresponsive to stimulation by fresh males, showing that the drug itself has an irreversibly debilitating effect directly upon female worms.

The anticancer drug procarbazine, known to cause testicular atrophy in human patients, induces an extreme but temporary involution of the testes in *S. mansoni* but has little effect on females (Basch and Clemens 1989).

In *S. mansoni*, Nollen (1975) implanted single worm pairs each with a radioactively labeled male plus 8–10 additional females into recipient hamsters, allowed them to interact for 10 days, removed the worms, and processed them for autoradiography. In 12 trials 4 inseminations were detected: 3 were with the female in copula and the other with a female not in copula, showing that some mate exchange can occur.

Parthenogensis

Parthenogenesis and asexual reproduction among parasitic platyhelminths are the subjects of an extensive review by Whitfield and Evans (1983), which reveals the complexity of the situation in both larval and adult digenetic trematodes.

Some female schistosomes are capable of normal development, sexual maturation, and production of viable eggs without pairing with a male. The best known example of unisexual maturation is that of *Schistosomatium douthitti,* in which numerous viable miracidia are regularly produced (Short 1952 a, b, 1957; Short and Menzel 1959). Unisexual females are shorter than paired females but their development is essentially similar. Basch and O'Toole (1982) found that *S. douthitti* females cultured in vitro from unisexual cercariae matured to a considerable degree in the absence of males, producing the inviable eggs characteristic of schistosomes grown in culture.

It was shown by Short (1952b) that uniparental miracidia of *S. douthitti* were infective to lymnaeid snails, although at a much lower rate than biparental miracidia. Subsequent cytological investigation (Short and Menzel 1959) revealed that in 168 parthenogenetic miracidial embryos, a haploid set of chromosomes occurred in 160 and a diploid set in 3; the remaining five were probably also diploid but were difficult to count. In contrast, of 153 miracidial embryos from bisexual mammalian infections the chromosome count was haploid in 4, diploid in 147, and triploid in 4. It was believed that diploidy in parthenogenetic eggs arose through meiotic irregularities, and that the occasional haploid embryo arising from normal bisexual infections developed parthoenogenetically despite the presence of male worms. This situation probably occurs from time to time in other schistosomes but has remained undetected.

In *Bilharziella polonica,* unisexual females develop normally and deposit eggs, normal in appearance but lacking a miracidium (Brumpt 1949). In both *S. douthitti* and *B. polonica* the female worms are distinctly flattened, broad, and fluke-like. The filiform unisexual females of *Heterobilharzia americana* were found by Armstrong (1965) to be shorter than paired females and to produce large numbers of eggs, all of which were infertile, in experimental infections in mice.

Within the genus *Schistosoma,* LeRoux (1951a) reported finding eggs in the liver, lungs, and intestinal wall of a mouse infected with unisexual female *S. mattheei.* Miracidia were hatched from some of these eggs, but exposed *Physopsis africana* snails did not become infected. Taylor et al. (1969) found that female *S. mattheei* produced about 500 eggs per mouse 17 weeks post unisexual infection. The eggs were small, distorted, and inviable. A subsequent study (Taylor 1971) showed that whereas female *S. mattheei* continued to mature for 71 weeks in the absence of males, no viable eggs were ever produced and by 88 weeks post infection egg production had apparently ceased. Van Rensburg and van Wyk (1981), working with *S. mattheei* in sheep, found no ova in unisexual females at 71 days, but did see some at 134 days. It is quite possible that different genetic stocks of *S. mattheei* were used by these authors.

In the case of *S. haematobium,* Sahba and Malek (1977a) reported abundant egg production by unisexual females of Iranian origin, with up to 5,622 eggs per gram of mouse liver at 5 months of infection, with a mean of 4.7 uterine eggs per female. Whereas some eggs appeared to contain a partly developed embryo, no miracidial hatching was detected. About 15% of unisexual female *S. mansoni* were found to develop vitelline glands and "intrauterine bodies" when maintained for more than 145 days in 6 of 11 guinea pigs (Paraense 1949). These bodies were abortive eggs which, while very polymorphic, usually showed the lateral spine characteristic of *S. mansoni.* This tendency toward parthenogenesis has not been noted in other studies on unisexual female *S. mansoni* and may be a peculiarity of that particular population from Rio de Janeiro, Brazil.

Experimentally, the classical studies of Vogel (1941a, b; 1942a, b) indicated that females of *S. mansoni, S. japonicum,* and *S. haematobium* coinfected with males of the other species in all combinations matured and produced some eggs typical of the maternal type, which were presumed to have developed parthenogenetically. Similarly, Taylor (1970), working with species complexes of African schistosomes, found that *S. mansoni* females will mature when coinfected with *S. rodhaini* males, and *S. rodhaini* and *S. mattheei* females develop with *S. mansoni* males, with production of normal eggs.

Short (1948a) and Armstrong (1965) coinfected mice with female *S. mansoni* and male *S. douthitti,* reporting pair formation and egg production. This relationship was explored further by Basch and Basch (1984), who found that many mixed pairs formed. The paired females were approximately the same size as those in unisexual infections, far smaller than females paired with *S. mansoni* males. Although the *S. douthitti* males possessed well developed testes, sperm were not found in their female partners. The *S. mansoni* females developed scanty vitelline glands and produced laterally spined eggs that yielded swimming miracidia infective to *Biomphalaria glabrata,* but not to the lymnaeid host of *S. douthitti.* Sporocysts arising from these miracidia were haploid and produced cercariae infective to mice. Mice exposed to cercariae from 1 of 10 individual shedding snails (not unimiracidially infected) all harbored female worms, and in 2 cases male worms were also present, indicating that 2 snails had each become infected by at least 1 female and 1 male-producing miracidium. The apparent 5F:1M sex ratio was unexplained. Parthenogenetically derived female cercariae in mice coinfected with either parthenogenetically-derived male or normal diploid male *S. mansoni* developed to large adults of normal appearance, whose eggs yielded diploid miracidia and subsequent generations of normal diploid schistosomes. Parthenogenetically derived *S. mansoni* females coinfected with *S. douthitti* males also paired and produced some eggs containing viable miracidia, which gave rise once again to haploid sporocysts. These observations confirmed previous suggestions that the stimuli for maturation in female *S. mansoni* is distinct from that for growth, and is independent of insemination and fertilization. It was concluded that both diploid and haploid *S. mansoni* females are capable either of parthenogenesis or of bisexual reproduction when appropriately stimulated.

In areas where several schistosome species coexist, the ability to develop parthenogenetically may represent a survival mechanism, by maintaining the germ plasm of the maternal species for some generations until a conspecific male appears for fertilization within a host (Basch and Basch 1984).

Unisexual Infections

Cort (1921b), referring to previous published and unpublished work by Japanese parasitologists, forwarded the hypothesis that sex in the schistosomes is determined in the fertilized egg and all the cercariae coming from a single miracidium are of the same sex.

> When all the individuals derived from the cercariae from a single snail were of the same sex it would follow that the infestation in this snail was from a single miracidium or two or more miracidia of the same sex. In those cases where both sexes came from the same snail, this snail must have been originally infested with two or more miracidia representing both sexes. (Cort 1921b)

This hypothesis has now become dogma in the sexual biology of schistosomes.

Numerous investigations have employed monomiracidially infected snail hosts in order to obtain cercarie of one or the other sex for animal infections, always confirming the original observation of Cort. A summary of recorded sex ratios resulting from such unimiracidial infections is given in Table 4–2. Unfortunately, it is not possible to determine the sex of cercariae by simple morphological or behavioral examination. The standard procedure for obtaining animals with a schistosome infection of one known sex still requires preliminary infection of a test animal and its sacrifice after schistosomular development to a distinguishable sex. During this interval it will be earnestly hoped that the source snail remains in good health. Various cytological and molecular techniques for discrimination between male and female cercariae were discussed earlier.

Unisexual Male Infections

The relative development of schistosomes in unisexual and bisexual infections has elicited much interest, but information is available only for the most frequently studied species. In general, male schistosomes are less altered in the absence of the opposite sex than are female worms. This was pointed out by Vogel (1941a), who showed that body growth and development of presumably West African male *S. mansoni* over a period of 5 months was identical in unisexual and bisexual infections. Using parasites of Puerto

TABLE 4–2 Sex Ratio of Infections from Single Miracidial Exposures

Species	Snail Host	Infection		Comment	Reference Note
		Male	Female		
S. mansoni	Biomphalaria glabrata	20	19		1
	B. glabrata	25	17		2
	B. glabrata	38	49		3
	B. glabrata	57	34		4
	B. pfeiferri	5			5
	—		"more"		6
S. rodhaini	B. sudanica	1	1		7
S. haematobium	Bulinus truncatus	7	4	Iran	8
	Bulinus globosus				
	Bu. t. rohlfsi				
	Bu. globosus, etc.	9	2		9
S. mattheei	Bu. globosus	22	8		10
	Bulinus	1	5		11
	Bu. globosus, etc.	6	9		12
S. bovis	Bu. truncatus	6	9		13
S. intercalatum	Bu. globosus	41.6%		Zaire	14
	Bu. forskalii	58.3%		Cameroon	15
S. japonicum	Oncomelania hupensis	12	13		16

B = Biomphalaria; Bu = Bulinus

Reference notes: 1, Vogel 1941b; 2, Taylor 1970; 3, Lancastre et al. 1984; 4, Liberatos 1987; 5, Vogel 1941b; 6, Rowntree and James 1977; 7, Taylor 1970; 8, Sahba and Malek 1977a; 9, Wright and Ross 1980; 10, Taylor 1970; 11, Taylor et al. 1969; 12, Wright and Ross 1980; 13, Taylor 1970; 14, Frandsen 1977; 15, Frandsen 1977; 16, Vogel 1941b.

Rican origin, Armstrong (1965) concluded that "The development of S. mansoni males in unisexual infections differed very little, if at all, from that in bisexual infections." In a Brazilian isolate of the same species, Zanotti et al. (1982) measured worms recovered 60 days post infection from 51 mice with male unisexual, and 30 mice with bisexual infections. The mean length of worms in the former group was 5.82 mm, and in the latter, 6.68, significant at the 1% level by t-test. The mean number of testes was 7.80 in unisexual and 8.13 in bisexual worms, significant at the 5% level.

Although morphological and developmental differences between unisexual and paired males are not so pronounced in S. mansoni, their distribution in the host and their physiologic responses are not the same. Standen (1953b) showed with an unspecified but probably Egyptian strain that paired males are more often in the mesenteric veins, and less often in the portal vein or liver than are unisexual males. Armstrong (1965) dropped mice into liquid nitrogen at intervals from 4 to 51 weeks post infection in order to determine the precise distribution of the schistosomes; the overwhelming majority (97%) of unisexual worms was in the liver and portal vein. Over a period of time Standen found 14.5% of unisexual males in the mesenteric veins, while Armstrong found almost none. It was suggested by Standen (1953b) that in all-male infections of S. mansoni, a moderate degree of migratory impulse

TABLE 4–3 Lengths of Unisexual and Paired Female *S. mansoni*

Origin/Host	Unisexual				Paired				Reference Note
	Number	Range	Mean	S.D.	Number	Range	Mean	S.D.	
Puerto Rico									
Mouse	20	2.6–4.3	3.7	0.44	20	7.9–13.8	9.26	3.36	1
Mouse	20	3.4–6.0	4.67	0.75	20	8.2–13.3	9.9	3.62	
Mouse		3.5–6.5	5.5[a]						2
Mouse			3.8[b]						3
			4.9[c]						
			5.5[d]						
Mouse		2.7–6.0	4.38[e]		13		7.26		4
Belo Horizonte, Brazil									
Mouse			3.48	0.21			5.61	1.9	5
West Africa?									
Mouse	22	1.6–3.6	2.66		13	1.6–3.4	2.52[f]		6
Mouse	25	2.7–4.5	3.44		6	6.2–8.3	7.46[g]		
Mouse	15	2.9–5.0	3.68		6	6.2–10.7	7.90[h]		

Footnotes: [a]Does not change between 40 and 80 days; [b]30–39 days postinfection (dpi); [c]90–99 dpi; [d]120–129 dpi; [e]8 weeks pi; [f]1 month pi; [g]3 months pi; [h]4 months pi.

Reference notes: 1, Basch and Basch 1984; 2, Armstrong 1965; 3, Shaw and Erasmus 1981; 4, Moore et al. 1954; 5, Zanotti et al. 1982b; 6, Vogel 1941a.

exists, which causes a variable proportion of the male worms to leave the liver and to occupy the portal and mesenteric veins. Standen believed that the migratory impulse in males is considerably stimulated by the presence of and pairing with females just sexually mature. As evidence for this view he pointed out that when hosts with established all-male infections were reinfected with female cercariae, migration of paired worms commenced only after the 28th day of female infection, although the males were physically capable of moving to the mesenteric veins at any time.

Regarding physiology, Tomosky et al. (1974) demonstrated that responses of unisexual male worms to dopamine, norepinephrine, and epinephrine were quite different from those of males paired with females.

In *S. japonicum* the differences beween males in unisexual and bisexual infections were explored quite early. Faust and Meleney (1924), in China, suggested that the absence or paucity of females was in some way correlated with lack of maturity in males. Shortly thereafter Severinghaus (1928), also in China, made an elaborate study of reproductive development in *S. japonicum*, finding that unisexual males matured more slowly, at about 75 days post infection, attaining almost the size of paired males. Thereafter, the single males were found to become smaller again, suggesting a sort of degeneration. Testes developed normally with fully mature sperm at 54 days post infection. In the same year, in Japan, Sagawa et al. (1928) noted that although the testes "gradually matured with the days" the size and density of testes in unisexual males was considerably inferior, and their growth was slow. The growth of other organs such as the gynecophoral canal was remarkable only in cases of bisexual parasitism. Vogel (1941a) measured 22 *S. japonicum* males 83 days old from unisexual mouse infections, finding a mean length of 5.08 mm; 7 males from conspecific pairs averaged 8 mm in length, demonstrating a clear retardation in growth in unpaired males of this species.

In *S. haematobium*, Sahba and Malek (1977a) compared dimensions of males from unisexual and bisexual infections, finding a mean length of 8.22 and 10.83 mm, respectively. Males acquired the normal number of testes, with production of spermatozoa.

In *Schistosomatium douthitti*, Short (1952a) found the development of males to be similar in unisexual and bisexual infections. Sperm was observed in unisexual males 14 days post infection, one day later than in the presence of females, and the lengths of both types of males were about the same.

Unisexual Female Infections

Little is known about female unisexual infections in most species of schistosomes. Most studies have employed either *S. mansoni* (Table 4–3) or *S. japonicum*, in both of which female growth and maturation are severely hindered in the absence of males. More than 60 years ago Suyemori (1922) noted that in *S. japonicum* from Taiwan, infections arising from single naturally infected snails were almost always of one sex only. In unisexual females, "even after the lapse of considerable time from infection, the worm retains the shape of the young fluke . . . only increasing a little in the length of the body." In Japan, Sagawa et al. (1928) made a similar observation, finding that the growth of ovaries and other female genital organs was very slight, and "it would not be too much to say that these female genital organs could only accomplish their growth in cases of bisexual parasitism." That the Chinese *S. japonicum* behaved the same was confirmed by Severinghaus (1928), whose "independent females" remained at about one fifth of normal body size and were lacking in most reproductive organs, having only the beginning of an ovary. Both Sagawa et al. and Severinghaus reinfected their animals (rabbits and Chinese hamsters, respectively) with male cercariae, reporting that maturation was then attained by the female worms. All three of these publications also stated that the excess females in unbalanced infections resembled unisexual female worms. However, Li (1958) found that in mice, worms in copula seemed to exert an inhibitory effect on extra unpaired females, which were even smaller than females in unisexual infections.

Many studies have been done with unisexual female *S. mansoni*. Although it is hazardous to compare studies on parasites and hosts of different origin and genetic composition, using a variety of techniques over a span of several decades, several general principles have emerged. The pioneering work in this species was that of Vogel (1941a), who showed that while paired females in mice grew from 2.52 to 8.25 mm in length between 1 and 5 months post infection, unisexual females remained within the range of 2.6 to 3.7 mm (group means), for as long as they were followed, up to 10 months post infection. At 1 month the anatomical distinctions between unisexual and paired females were not complete: the ovary was visible as a small spiral structure and rows of cells indicating the anlage of the vitelline gland. The unisexual females remained in this state, whether examined at 2, 4, or 10 months, while in the paired females the reproductive system was fully functional by 38 days post infection, demonstrating sperm in the oviduct and an egg in the ootype. A similar study, with detailed description of reproductive development, was done by Moore et al. (1954). These authors started with unisexual *S. mansoni* of Puerto Rican origin 11 weeks post infection, exposing each female-infected mouse to 150 male cercariae. Eight weeks later, paired females averaged 6.36 mm long, while unisexual controls remained at 3.9 mm. In these worms the fundament of the vitelline gland was visible as small clumps of cells adjacent to the cecum, with a duct visible only in well-differentiated specimens. The ovary was of a loose, undulant shape, containing small, compact, uniform oogonial cells. The ootype and uterus were completely developed. Similar observations were made by Armstrong (1965), who determined that in Puerto Rican *S. mansoni* the unisexual female stops development about 40 days post infection, when it resembles the stage described by Moore et al. Vitelline granules and small, malformed eggs were found in the uteri of about 5% of unisexual females (average length, 5.5 mm) 40 to 80 days post infection in the series of Armstrong.

The most detailed comparative study of the reproductive system in unisexual and mature female *S. mansoni* was done by Erasmus (1973), using both light and electron microscopy. Two main morphological differences were determined. First, in unisexual females the ovary, although smaller than normal, was capable of producing ova, but these were devoid of cortical granules. While Golgi complexes were able to synthesize some material, the product was not properly organized. Second, the vitelline cells in unisexual females were almost completely unable to mature. Both differences were found to be "genuinely related" to the absence of a male. A subsequent study (Shaw and Erasmus 1981) showed that the state of development of females from single sex infections is not as precisely defined as had been suggested. At 30 days post infection about 5% of females possessed scattered groups of vitelline cells. This was found to vary considerably among unisexual females within and between mouse hosts, the proportion of worms possessing mature vitelline cells generally increasing with age. No relation was found with worm number per host, sex of host, or length of the worm. The degree of development was similar at 30 or 200 days post infection, and a small percentage of females had eggshell material within the ootype. The role of the male was interpreted as accelerating and coordinating, but not necessarily initiating development. A special study of vitelline cell development in unsexual female *S. mansoni* was done by Shaw (1987). In unisexual infections the state of development achieved by *S. bovis* and *S. rodhaini* is similar to that found in *S. mansoni* (Taylor 1971).

Even more than for male worms, the distribution in the host of females depends on their state of pairing. Standen (1953b) pointed out that females in single-sex infections were never found in the mesenteric veins, remaining primarily in the liver and main portal vein. The reason for this is that the female appears physically unable to make the journey unassisted, depending upon the larger, more muscular male to carry her into the mesenteric venules. Although this seems to be so, there is no experimental evidence to confirm it.

Many authors have remarked on the difference in intestinal cecal contents between unisexual and paired female worms: the former are pale or brownish-granular in nature, indicating either that the worms do not feed well (i.e., are malnourished or starving), or that their enzymatic digestive apparatus is not adequately functional. Smith and Chappel (1990) mentioned that "elderly spinsters" of *S. mansoni* more than 12 months old "contain within their digestive tract significant quantities of haematin." Perhaps this trait is genetically controlled and varies from one isolate to another.

Sexually mature and immature (not necessarily unisexual) female *S. mansoni* were analyzed for protein patterns on 2-dimensional gel electrophoresis by Braga et al. (1989). In comparison with immature and mature males, which were the same, substantial differences were identified between immature and mature females, each of which had three proteins not present in the other. Further evidence for physiological differences between unisexual and paired *S. mansoni* was provided by Standen (1955), who showed that females in single-sex infections were unaffected by the drug 1:7-bis-(p-dimethylaminophenoxy)heptane, but went on to maturity when the hosts were later reinfected with male cercariae. When the males were 32–49 days old, the females became susceptible and were killed by the drug.

Maturation in unisexual *S. mattheei* has been noted by several authors, as described in the section on parthenogenesis (page 154), but the best documented case of complete development of unisexual female schistosomes is in *S. douthitti*, well described by Short (1952a and other publications). About 25% of such females migrate from the liver to the mesenteric venules in this species. Their well-known propensity to mature in the absence of male companions has its limitations: at 40 days after infection females from mixed infections attained about 3 mm in length, while unisexual females were about 2 mm. After a year or more, both groups were essentially the same length (Short 1952a).

Although also discussed elsewhere, mention should be made here of the experimental use of unisexual female schistosomes as sensitive indicators of maturational stimuli. Among the numerous studies those of Shaw at al. (1977) and Popiel and Basch (1984b) may be cited as examples. The use of fast red B salt (Johri and Smyth 1956) provides a simple means of detection of vitelline cell development, and the method of sporocyst cloning by surgical transplantation, although tedious, yields a continuing supply of genetically uniform material.

Significant differences in antigenic composition between *S. mansoni* females from unisexually and bisexually infected hamsters were reported by Rotmans and Burgers (1984).

Hermaphroditism

Having descended from monoecious ancestors, the schistosomes may be expected to demonstrate hermaphroditism under the proper circumstances, and indeed individuals having some characteristics of both sexes have often been reported. The primary sex is always expressed phenotypically in any individual. It is likely that over the course of evolution the genetic information needed to assemble and operate the reproductive apparatus of the secondary sex has suffered such degradation that functional hermaphroditism cannot now occur. The degree to which partial hermaphroditism is found in natural populations is poorly known, even for the intensely studied species, and no data are available for most of the genera parasitizing birds and wild mammals. Most earlier reports of hermaphroditism in schistosomes were summarized by Adam (1960).

Hermaphroditism in Female Schistosomes

The propensity of female schistosomes to demonstrate male characteristics appears to be much less than that of males to become partially feminized, according to published reports, but the reason for this difference is not clear. It may be that selection pressures towards greater reproductive productivity have acted more harshly upon the female genome, weeding out nonfunctional vestigial genes associated with maleness. Lagrange (1958), studying the Vogel (West African?) strain of *S. mansoni,* described unisexual female infections in 10 mice infected with cercariae shed from a single snail (possibly, but not necessarily unimiracidially infected). In nine of these mice normal undeveloped females were found, but some individuals also possessed male characteristics. The same material was restudied by Adam (1960), who stained and sectioned some of the specimens, describing them in greater detail. The seven "andromorph" females were long and slender anteriorly and possessed a gynecophoral canal in the body region posterior to the ventral sucker; dorsally, the body was covered with spinous papillae typical of male *S. mansoni.* Female reproductive structures were present, but there was no trace of testes. Two of these females possessed abortive eggs in the uterus, and in one case a normal female was found within the gynecophoral canal of an andromorph. All sectioned andromorphs had a histologically normal female reproductive system. A similar situation had been described earlier by LeRoux (1951a) in the case of *S. mattheei.* Since all of the reported andromorph females arose in cercariae from a single snail, it seems possible that a genetic alteration, perhaps a mutation in a parental spermatocyte or oocyte, led to expression of normally suppressed remnants of the male-associated genome.

In China, Ho (1962) described and illustrated (Fig. 4–2) a hermaphroditic female *S. japonicum* that had all normal female characters, including 140 eggs in the uterus, but also possessed 7 testes posterior to the ventral sucker. No seminal vesicle or male genital pore was found.

A possible effect induced by hormone administration was described by Robinson (1957a), who found 667 males, 434 females, and 240 "male-like hermaphrodites" (not further described, but suggested to be "reversed females") in the next generation of adult *S. mansoni* after testosterone treatment of mouse hosts. However, the same author reported later (Robinson 1957b) that he could not consistently confirm these results.

A female *Heterobilharzia americana* was illustrated by Sahba and Malek (1977a), with a normal female reproductive system plus 7 testes in the posterior part of the body. The most extensive report of hermaphroditism in female schistosomes is that of Short (1951), who found 21 of 1,350 female *Schistosomatium douthitti,* from both bisexual and unisexual infections, with testicular follicles. These follicles, lying posterior to the ovary between the intestinal ceca, were immature and lacked efferent ducts. In females 14 days or older, cells indistinguishable from primary spermatocytes were

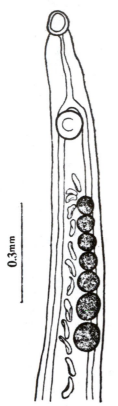

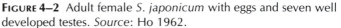

FIGURE 4–2 Adult female *S. japonicum* with eggs and seven well developed testes. *Source*: Ho 1962.

found in these structures, arranged in groups of eight as characteristically found in the testes of males. Such observations support the concept that retention of the male-specific portion of the hermaphroditic ancestral genome, although partial, is more extensive in *Schistosomatium* than in *Schistosoma*.

Hermaphroditism in Male Schistosomes

Hermaphroditism of male *S. mansoni* of West African origin was well documented by Vogel (1947) (Fig. 4–3). Numerous males developing in guinea pigs were found to have rudimentary female structures, corresponding to one or even several ovaries, developing in the midbody between the intestinal branches. In the better developed of these structures numerous cells were seen to resemble oocytes; in some (17 specimens) a well-defined duct resembling a uterus led forward almost to the acetabular region, where it ended blindly. However, most males with ovaries lacked the duct. Additionally, some males possessed strands of cells grossly and histologically resembling vitelline glands in the posterior body where such structures are normally found in females. Vogel (1947) pointed out that although the testes are as a rule normally developed in male worms with ovaries, in some individuals the female organs had developed at the expense of the male glands, with

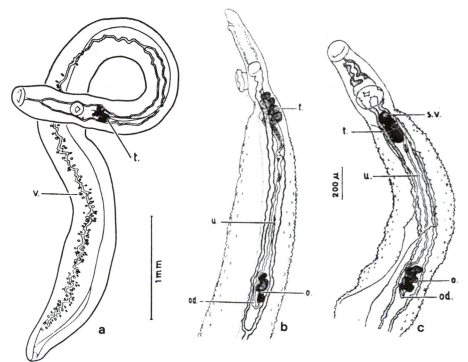

FIGURE 4–3 Degrees of hermaphroditism in male *S. mansoni*. a, Extensive vitelline cell development; b, c, Presence of ovaries and female genital ducts in addition to the testes. From guinea pigs. o, ovary; od, oviduct; s.v., seminal vesicle; t, testis; u, uterus; v, vitelline follicles. *Source:* Vogel 1947.

a reduction in number or state of development of the testes. In extreme cases, the testes were almost completely lacking. In guinea pigs exposed to both sexes of cercariae, 4.3% of males had an ovary, but in guinea pigs unisexually infected, 43.4% of males were hermaphroditic. European hamsters (*Cricetus cricetus*) exposed only to male cercariae produced adult worms 40% of which possessed an ovary, but only 1 of 51 males (2%) from bisexual infections showed this trait. A single snail shedding male cercariae was used to infect both mice and guinea pigs: the former had no hermaphroditic males, while in the latter "almost half" the worms possessed an ovary. Rabbits also favored development of hermaphroditism, appearing to harbor the more extreme cases.

The particular genetic constitution of the isolate, the species and possibly the genotype of the host, and perhaps the presence and number of female worms are of crucial significance in the development of hermaphroditism in *S. mansoni*. A similar phenomenon was never observed in comparable infections with *S. haematobium* and *S. japonicum* (Vogel 1947).

Vogel's results were confirmed by Lagrange and Scheecqmans (1949), who apparently obtained their *S. mansoni* from the same source and reported that 80% of their males were hermaphroditic in guinea pigs. Gönnert

(1949b), using Vogel's isolate of *S. mansoni* in mice, found no true hermaphrodites but did report and illustrate vitelline material with well developed vitelline cells and a rudimentary vitelline duct, beginning shortly behind the testes and extending posteriad between the intestinal ceca until near the posterior end. Buttner (1950, 1951), working with the same line of *S. mansoni*, found all of 104 males from guinea pigs with a well differentiated ovary. She reported, moreover, that the development of an ovary was accompanied by a reduction in the number of testes, with a corresponding inhibition of the male sexual apparatus causing the males to be unsuitable for fertilization of the female worm. No hermaphroditic worms were recovered from mice or rabbits. Not only is the species of host of significance, but the genetic makeup of the *S. mansoni* is also critical, for Buttner was unable to demonstrate any hermaphroditism, even in guinea pigs, in worms of South American origin. In a comparative study of *S. mansoni* from various areas, Saoud (1965), using hamsters and mice, found hermaphroditism (otherwise undescribed) in 2% of males originating from Mwanza, Tanzania; 4% of males of Egyptian origin; and none in males of a Puerto Rican isolate.

A year after the report of Vogel (1947), Short (1948a) discovered hermaphroditic male *S. mansoni* in commercially prepared teaching specimens of unknown animal origin from Puerto Rico. This material agreed in general with Vogel's descriptions, but some males, in addition to well developed ovaries with typical oocytes, had oviducts, ootypes, Mehlis' glands, short uteri, main vitelline ducts, and vitelline glands in different stages of completeness. Three males showed a female genital pore, which united the male and female ducts in what was considered a common genital pore. Most oviducts contained granular materials and vitelline cells were found in the duct of one worm. In Brazil, Paraense (1949) exposed guinea pigs to *S. mansoni* cercariae from snails collected on the grounds of the Instituto Oswaldo Cruz in Rio de Janeiro. He noted the presence of a rudimentary ovary in 23 of 78 and in 22 of 88 male worms from two animals, but none in 11 other guinea pigs. Apparently normal testes were found in the hermaphroditic worms. Seemingly unaware of this work in their country, Ruiz and Coelho (1952) had infected guinea pigs with *S. mansoni* cercariae from *B. glabrata* collected in the states of São Paulo and Minas Gerais. Of 770 male worms, 48% possessed ovaries identical to those described by Vogel. A large number of these worms also possessed vitelline follicles along the cecal branches. The number of testes was not reduced, and the presence of secondary hermaphroditism did not interfere with pairing or reproduction, as female worms developed normally and possessed eggs in their uteri. Vitelline cells in males, of Puerto Rican origin, were reported once again by Shaw and Erasmus (1982) who found Fast Red B positive material in worms 30 to 100 days post infection. Electron microscopy revealed cells with characteristics of mature stage 4 vitelline cells, such as vitelline and lipid droplets, glycogen and calcareous corpuscles as reported from mature female worms by Erasmus (1975b) and Erasmus and Popiel (1980). However, nothing resembling vitelline lobules was found.

It was believed by LeRoux (1951b) that male *S. mansoni* showed a tendency to hermaphroditism only after several previous passages in which the number of females exceeded that of males; in subsequent passages up to 100% of males were said to show this trait. He thought also that sexually abnormal males may be influenced by such factors as the temperature of the water and nutritional state of the intermediate host snails.

In species of *Schistosoma* other than *S. mansoni*, hermaphroditism was found in males of a Taiwan isolate of *S. japonicum* by Moore (1955). A single male possessed an ovary and oviduct, but no vitelline structures; testes and sperm appeared normal. In a subsequent more complete study, Hsü and Hsü (1957) investigated 4,064 *S. japonicum* males of Formosan, Chinese, Philippine, and Japanese origin. In bisexual infections, 6 males with ovaries were observed (3 from a dog, 2 from mice, 1 from a hamster), all of Chinese origin (1.7%), none in other worms. Four of the 6 worms had only a rudimentary ovary. Guinea pigs, hamsters, and rabbits were infected unisexually with males of the Japanese and Formosan isolates, but no hermaphroditism was observed among 1,517 such males. Hermaphroditism in *S. japonicum* has also been reported by Chien and Liu (1960).

In *S. haematobium*, isolated oocytes were observed near the testes of 12.9% of paired males examined by Deters and Nollen (1976). The cells resembled primary oocytes of the posterior part of the ovary of the female and incorporated ^{3}H-thymidine into the cytoplasm but not the nucleus. The prevalence of testicular oocytes was less in older males.

During a first passage of *S. bovis* from Sardinia in mice, Saoud (1964) found 2 of 6 males in a bisexual infection with ovaries similar to those described for other species. This author reviewed a previous collection of *S. bovis* from Sardinia and found about 35% of males also with an ovary; however, none of over 200 male *S. bovis* from Kenya showed such a structure. In a large series of this species from Israel, Lengy (1962b) was unable to detect any hermaphroditism.

Little has been reported about hermaphroditism in other genera of schistosomes. Ulmer and vandeVusse (1970) have observed that all males of *Dendritobilharzia pulverulenta* from 21 host species of anatid birds are "hermaphroditic" insofar as they have a distinct median ventral pit in precisely the position occupied by the genital pore in females. No trace of ducts or female organs was found either in whole mounts or sections of male worms.

Homosexual Pairing

Although male-male homosexual pairing has often been noted in schistosomes, a single report has been found describing female-female pairs. As described by Adam (1960), a normal female of *S. mansoni* was embraced by an andromorph female. Interestingly, the clasped worm was said to have its genital structures completely developed, suggesting that the andromorph fe-

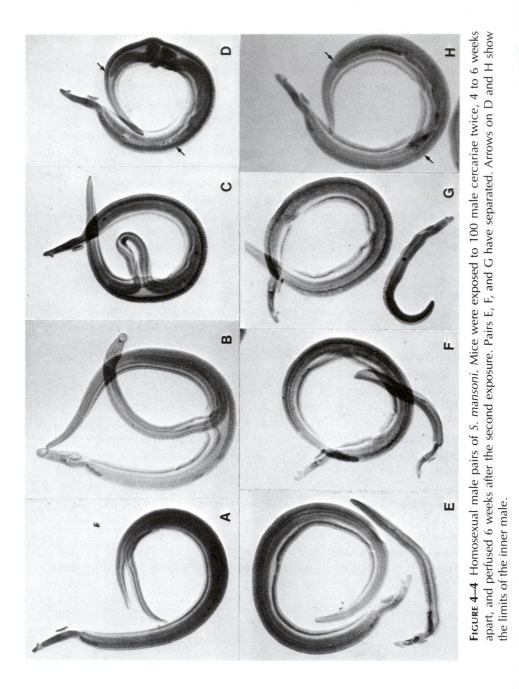

Figure 4–4 Homosexual male pairs of *S. mansoni*. Mice were exposed to 100 male cercariae twice, 4 to 6 weeks apart, and perfused 6 weeks after the second exposure. Pairs E, F, and G have separated. Arrows on D and H show the limits of the inner male.

male was capable of providing the necessary maturational stimulus as if it were a male.

In males, most reports have dealt with *S. mansoni.* Giovannola (1936) found 3 male pairs in a unisexual infection in a rabbit derived from ova from a human case in Puerto Rico. In each pair the male being clasped was the smaller. An identical observation was made by Paraense and Malheiros Santos (1949), who found 3 male pairs in 2 mice infected with worms of Brazilian (Belo Horizonte) origin. Vogel (1947) repeatedly encountered male pairs of *S. mansoni* and also of *S. japonicum* in his experimental animals, also observing that as a rule the larger male enclosed a smaller one. In two stained male pairs, the smaller male was hermaphroditic, suggesting that residence in the gynecophoral canal converts a male into a hermaphrodite, inciting the development of latent female genital anlage. This condition was never seen in males of *S. japonicum* or *S. haematobium.* The partners of male couples separated more readily than those of heterosexual pairs. Vogel (1947) presumed that "most probably the loose contact of male partners results in frequent separations and reunions within the blood vessels of the living host." Using the same parasite line as Vogel and Gönnert, deCarneri (1963) reported 4 male pairs among 218 worms in mice with unisexual male infection, and Lagrange and Scheecqmans (1949) found 4 male pairs, 2 of which were in the reversed position, in guinea pigs.

Small, sexually immature male *S. mansoni* were occasionally found in the gynecophoral canal of larger males by Armstrong (1965) 30 to 50 days post infection, but only where there were few or no females. Male pairs were located primarily in the liver and portal vein, not in the mesenteric veins. When mice with established male infections were again exposed to mice cercariae, the older ones apparently embraced newly arrived and sexually undifferentiated males as the latter reached the liver and, by so doing, inhibited the development of the younger worms. Similarly, small males of *Heterobilharzia americana* were recovered rather frequently from within the gynecophoral canal of larger males of the same species. In mixed infections in mice, homosexual cross-pairing was common: *Schistosomatium douthitti* males embraced by *H. americana* were often poorly developed both sexually and somatically, while those held by *S. mansoni* males were undersized, as were *S. mansoni* males held by *H. americana* males (Armstrong 1965).

In India, Fairley et al. (1930) found immature schistosomes enveloped in the gynecophoral canal of more mature males of *S. spindale* within the liver of guinea pigs. Although the sex of the enfolded worms was not always determinable with certainty, they appeared to be maldeveloped males. In the same country, Rao (1938) described a similar phenomenon in guinea pigs infected with Cercariae Indicae XXX, presumed to be *S. nasalis.*

Instances of pairing of two male *S. douthitti* were found rather frequently by Short (1952a) both in unisexual infections and in bisexual infections, especially when males were in the majority. When both worms were fully grown, a part of one was usually held by the other, but sometimes pairs consisted of a smaller male almost completely enclosed by a larger one.

Homosexual Male Pairs to Induce Hermaphroditism

On the assumption that male schistosomes retain fragments of an ancestral female part of the genome from their hermaphroditic forbears, we infected mice with only male cercariae and then, after varying intervals, infected the same mice again with male cercariae of *S. mansoni* (Basch and Gupta 1988). The intention was to have the older males clasp the younger ones. If the gynecophoral canal environment contains factors that can cause the expression of female-specific genes, then a male worm carrying some of those genes might be induced to show some hermaphroditic features. Although up to 30% of males were paired (Fig. 4–4), no sign of ovaries or other female reproductive structures was seen in the clasped males.

Functions of the Male

Basch (1990) speculated that:

1. Under normal circumstances, conspecific male schistosomes inseminate females and fertilize their oocytes. However, this service may be contravened in some species and appears to be optional in others.
2. When the host can acquire male and female worms from different genetic pools, fertilization certainly increases genetic diversity.
3. In the process of sexual differentiation, early schistosomes contracted to preserve the biological integrity of their species through an unequal division of the functions of their hermaphroditic forbears. At least in the genus *Schistosoma,* females lost the determinants for maleness, plus most muscle and parenchymal tissue in exchange for greater reproductive efficiency. By way of compensation, those structures were increased in males to serve not only their own needs but those of their female partners as well; that is, male schistosomes assured their survival, in the main, not by donating sperm but by offering muscle.
4. Male *Schistosoma* are responsible for the physical transportation of females from the portal sinus to sites from which their eggs can exit the host efficiently. As this distance can be substantial in larger hosts such as humans and big herbivores, it is to their advantage as efficient tractors for males to have as much beef as possible within the constraints of life in the mesenteric veins.
5. The same muscularity of male schistosomes permits them to assist the weaker females physically to pump host blood into their intestine by a massaging action of the muscular walls of the gynecophoral canal. Her growth and development ensue as a function of nutrition and not as a result of any mysterious growth factor emanating from the male.

In some other genera such as *Schistosomatium,* males are relatively smaller and less muscular than in *Schistosoma,* and females relatively larger and more robust. Female *Schistosomatium* have sufficient body and pharyngeal musculature to transport themselves to the mesenteric veins of their rodent hosts, and to ingest sufficient blood for growth and development. Female *Schistosomatium* have retained some ancestral male genes, as shown by their tendency to become hermaphroditic.

5

Relations with Snail Hosts

Stages Beyond the Mammal

The Miracidium

All of the reproductive effort of adult schistosomes is directed toward the production of miracidia, the fragile, indispensable bridge between the mammalian and molluscan hosts. The miracidium has been described in great detail (more than 400 pages!) by Ottolina (1957), and also by Schutte (1974a) and Pan (1980).

Development. Earliest miracidial embryology in *S. douthitti* was detailed by Nez and Short (1957); presumably these events are similar in other schistosomes. Just after egg deposition, the oocyte nuclear membrane disappears, bivalents become visible, and diakinesis occurs. The spermatozoan nucleus lies at the periphery of the first metaphase plate and during anaphase 1 is near the chromosomes of the secondary oocyte. Two polar bodies are formed: the first separates from the ovum and may divide; the second remains nearby. The sperm nucleus now becomes vesicular and the male and female pronuclei probably undergo prophase separately with all chromosomes mingling first on the metaphase plate. The first cleavage gives a small daughter cell and a larger one, which soon divides to form a 3-cell embryo that then goes on to form the miracidium.

In *S. japonicum,* differentiation of the embryo from the fertilized ovum to ciliated miracidium in the tissues of mice and rabbits requires 11 to 12 days (Ho and Yang 1979). The course of development was divided by these authors into four morphologically and histochemically distinguishable stages: (1) single-cell stage, in which the vitelline cells are rich in lipids; (2) cell-cleavage and (3) organ formation stages, when lipid decreases and nucleic acids and proteins increase markedly. The reaction for polysaccharides develops in the space between the embryo and eggshell; (4) mature miracidium. Here nucleic acids are concentrated in the neural mass and germinal cells. Lipids have almost disappeared, and abundant carbohydrates are found in the miracidium. A complex of chemical components consisting of neutral polysaccharides, proteins, and enzymes is identifiable in the cephalic glands. Secretions of these and the apical glands are now extruded through the eggshell, leading to complex in vivo circumoval precipitates. These appear as early as 41 days postinfection in the mouse.

Blair (1966b) found that the schistosome miracidium at first always points in the direction opposite to the spine. As it develops and begins to wriggle within the eggshell, this directionality is lost, so that in 1,000 *S. haematobium* eggs examined the miracidium pointed towards the spine in 508 and away from the spine in 492. For *S. mansoni* the corresponding figures were 234 and 266 (Fig. 5–1).

Hatching

Once the egg is passed from the body of the host, the miracidium is capable of hatching within a few minutes of immersion in water appropriate to the environmental requirements of the miracidium. A useful brief review of the hatching of schistosome eggs was presented by Xu and Dresden (1990). For the egg of *S. mansoni*, and presumably for those of other species, actual hatching is preceded by a period of activation studied in detail by Becker

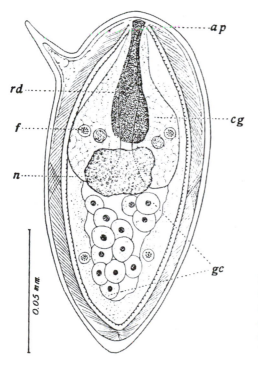

FIGURE 5–1 Miracidium of *S. mansoni* within the egg. ap, anterior papilla; cg, cephalic gland; f, flame cell; gc, germ cell; n, central nervous mass; rd, rudimentary digestive sac. *Source:* Cort 1919.

(1973, 1977). Osmosis-controlled processes include flame cell function in the protonephridial tubules, muscular contraction, and ciliary movement. These can be turned on and off several times by changing the salt concentration in the surrounding water. Egg volume increases from 205,000 to 270,000 cubic micrometers, and density declines from 1.1148 to 1.0984 upon immersion in distilled water, and an increase in water content of approximately 18% is sufficient to obtain activation in 90% of eggs (Becker 1977). In contrast, Samuelson et al. (1984) found that neither miracidia nor eggs swelled prior to hatching, which began with active beating of the miracidial cilia when the egg was placed in artificial pond water. These authors recorded and analyzed the hatching of *S. mansoni* eggs with the aid of video photomicroscopy. Kusel (1970) believed that osmotic forces are not the only ones that function during hatching. Incubation of living eggs in 1M glycerol followed by 0.9% saline resulted in hatching in 75%, but only 5% of those preincubated in 1M sucrose eventually hatched. In dead eggs similarly treated, 25% of each group "hatched" a dead miracidium. It was postulated that glycerol extracts, solubilizes, or releases a factor that may be a hatching enzyme.

Support for the hypothesis that osmotic pressure is not the exclusive trigger for hatching has come from Matsuda et al. (1983), who found that the antischistosomal drug praziquantel stimulated hatching in eggs of *S. japonicum* even under conditions of high osmotic pressure. Working with a Kenyan isolate of *S. mansoni,* Katsumata et al. (1988, 1989) concluded that a calcium-mediated process pays an essential role in hatching. Hatching was inhibited in a dose-dependent manner by the calcium channel blocker diltiazem and by the calmodulin antagonist W-7, suggesting calcium ion and calmodulin dependence of the hatching process.

Various authors, such as Abou Senna and Basaly (1964), Bair and Etges (1973), Magath and Mathieson (1946), and Upatham (1972a, b) have studied the effects of environmental conditions such as light, oxygen tension, temperature, and reducing environment upon hatching of schistosome eggs. The process of hatching has been observed in *S. mansoni* and *S. haematobium* by LoVerde (1976), who described a longitudinal fracture plane on the broad surface of the eggshell, from which the miracidium emerges. Similar observations and scanning electron micrographs were published later for *S. mansoni, S. haematobium,* and *S. mattheei* by Higgins-Opitz and Evers (1983), apparently unaware of the earlier work. The sequence of events during hatching were described as (1) activation of a trigger mechanism that creates the orifice when the egg is placed in water; and (2) emergence of the miracidium, with rupture of the surrounding membrane. The exact mechanism of formation of the orifice has not been described.

The Swimming Miracidium

Although easily mistaken for a ciliated protozoan, the miracidium is a complex multicellular animal specialized for the sole purpose of finding and penetrating a suitable snail host. All of our knowledge of miracidia comes from the laboratory. In nature, it is likely that the vast majority of miracidia fail to accomplish this mission and are eaten by micropredators or die and disintegrate within minutes to a few hours after hatching.

Early studies by light microscopy (Fig. 5–2) elucidated most of the major features of miracidia. Scanning electron microscopy of miracidia has been performed by various authors, including Køie and Frandsen (1976), who placed intact miracidia in an ultrasonic cleaner to produce "bald" miracidia by shaking off the cilia. Among other structures revealed in this way were epithelial cells, intercellular ridges, lateral processes, and a variety of simple and multiciliated pits that represent nerve endings: all disappear within a short time during the miracidial-sporocyst transformation. A pair of excretory pores at the level of the third row of epithelial cells probably remains in the sporocyst (Køie and Frandsen 1976). The ultrastructure of the miracidium of *S. mansoni* has been described most elegantly by Pan (1980) (Fig. 5–3), whose publication remains the gold standard for miracidial morphology.

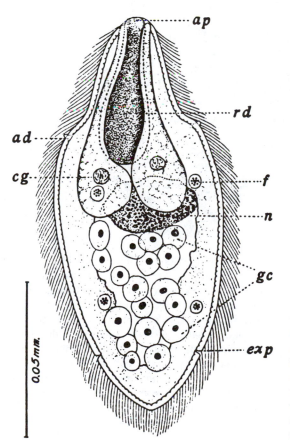

FIGURE 5–2 Miracidium of *S. mansoni*. ad, anterior ducts; ap, anterior papilla; cg, cephalic gland; exp, excretory pore; f, flame cell; gc, germ cell; n, central nervous system; rd, rudimentary digestive sac. *Source:* Cort 1919.

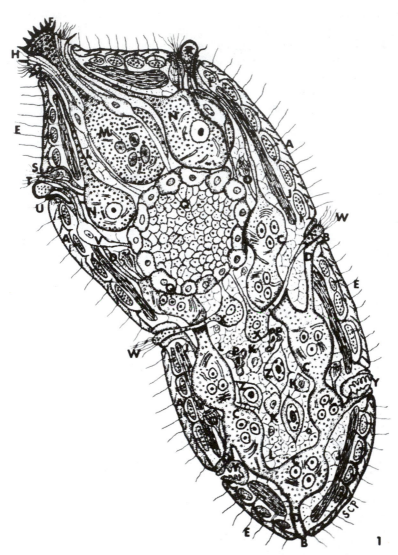

Figure 5–3 Schematic reconstruction of cellular architecture of the miracidium of *S. mansoni*. A, epidermal plate; B, epidermal ridge; C, ridge cyton; E, cilium and its rootlet; F, terebratorium and the profile of cytoplasmic expansions; G, multiciliated, deeppit sensory papilla; H, unciliated sensory papilla; I, outer circular muscle fiber; J, inner longitudinal muscle fiber; K, interstitial cyton; L, processes of interstitial cells; M, apical gland and secretory duct; N, lateral gland and secretory duct; O, flame cell with excretory tubule; P, cyton of common excretory tubule; Q, neural mass with peripheral ganglia; R, lateral papilla; S, multiciliated saccular sensory organelle; T, multiciliated shallow-pit sensory papilla; U, unciliated sensory papilla; V, perikaryon of the neuron to lateral papilla; W, multiciliated sensory papilla; X, perikaryon of the multiciliated sensory papilla; Y, excretory vesicle; Z, germ cell. *Source:* Pan 1980.

Schistosome-Snail Specificity

It is likely that any rich aquatic environment, freshwater or marine, will be inhabited by several species of mollusks that serve as hosts for larval trematodes. Resident and transitory vertebrates—fish, amphibians (in fresh water), reptiles, birds, and mammals—within, atop, and around the water body will introduce through their dejecta a continuous sprinkling of trematode eggs, from which every emergent miracidium will have at least potential access to each individual mollusk. Nevertheless, it is found invariably that year after year each host species is infected only with certain larval trematode species and never with others. How is such specificity controlled?

Behavior; Host-finding

The behavior of miracidia has been reviewed by Jourdane and Théron (1987). Samuelson et al. (1984) found that miracidia of *S. mansoni* swam at 2.5 mm/sec in artificial pond water and at 0.1 mm/sec in saline at 22C. Many authors have described positive phototaxis and negative geotaxis, or the reverse, in various schistosome miracidia with little rigorous definition or experimental evidence. Although a number of presumed sensory structures (Fig. 5–4) have been identified on miracidia, we have insufficient information about what miracidia see, smell, taste, feel, or think, nor exactly how they locate or recognize a potential host. In his review, Saladin (1979) commented that it "is regrettable that our comprehension of miracidial host-finding is not more commensurate with the effort that has been expended on the subject."

For species with swimming miracidia, including the schistosomes, one possible host restricting mechanism is miracidial behavior: swimming in microenvironments in which the most suitable hosts are most likely to occur, identifying and being attracted to acceptable candidate hosts, and rejecting others that they do not like. A second, non-mutually exclusive, scenario will have miracidia penetrating any available mollusk indiscriminately, including at least some individuals in which, for one reason or another, they are unable to develop. Those passively-transmitted trematode species whose eggs hatch only after they are eaten will naturally have little control over host selection.

A field-based study utilizing laboratory-raised miracidia and snails was carried out by Upatham (1973b) in "artificial banana drains" and natural streams on the island of St. Lucia.

> Miracidia located and infected snails at distances of 9.14 and 97.54 m horizontally in standing and running waters respectively. In running water no infection occurred above a velocity of 13.11 cm/sec. In both types of habitat infection rates in snails increased with increasing levels of miracidia but decreased as the location of caged snails moved away from the miracidial point of entry. . . . only a small percentage of miracidia succeeded in locating and penetrating snails (6.8–13.7% and 1.4–6.2% in standing and running waters respectively). (Upatham 1973b)

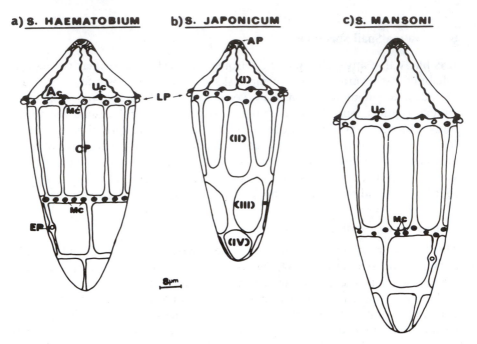

a) S. HAEMATOBIUM   **b) S. JAPONICUM**   **c) S. MANSONI**

FIGURE 5–4 Schematic drawing of the miracidia of *Schistosoma haematobium, S. japonicum,* and *S. mansoni.* The relative sizes, the shape of the epidermal plates, and the distribution of the sensory receptors are shown. Ac, aciliated sensory receptor; AP, apical papilla or terebratorium; CP, ciliated epidermal plate; EP, excretory pore; LP, lateral papilla; Mc, multiciliated sensory receptor; UC, unciliated sensory receptor; I, II, III, IV, order of tiers of ciliated plates. *Source:* Elku-Natey et al., 1985.

The controversial question of chemotactic host-finding capabilities of the swimming miracidium has been considered by Saladin (1979), who reviewed 176 published miracidium-snail combinations plus combinations of miracidia with other invertebrates, plants, and other objects. There is little evidence that miracidia are capable of identifying appropriate host individuals before attempting to penetrate them. Indeed, Saladin referred to laboratory trials in which miracidia of *Schistosoma mansoni* were incapable of discriminating among seven snail species, only one of which was an actual intermediate host. More than 60 organic chemical compounds including amino acids, as well as inorganic ions such as Ca^{++} and Mg^{++}, have proven attractive to miracidia of *S. mansoni* and of other species in the laboratory.

Many authors have shown that the miracidia of *S. mansoni* accumulate in snail-conditioned water (SCW). Samuelson et al. (1984) prepared SCW by leaving 10 large *Biomphalaria glabrata* in 10 ml of water for 3 hours at room temperature. By videotape analysis of swimming movements, they found that the relative number of miracidia within a freshly-placed spot of SCW increased 3-fold within 20 seconds, primarily because of a major increase in the turning rate of the miracidia, which reduced the SCW exit rate and kept them within the SCW spot. Dissous et al. (1986) have described a glycoprotein of 80 kDa molecular weight released by *B. glabrata* and capable of stimulating the metabolism of miracidia of *S. mansoni*.

From an evolutionary viewpoint, the capability for a miracidium to penetrate the "wrong" host, as described previously for the presumed intermediate host switch of the *S. japonicum* ancestor, represents an important device by which a phenotypically preadapted trematode larva may unlock the barrier to a whole new range of hosts, and a consequent burst of evolutionary radiation.

Relations with the Molluscan Host

The relations between schistosomes and their snail hosts are certainly as complex as those with their vertebrate hosts, but less well known to most parasitologists. For those unfamiliar with the morphology of planorbid snails, the anatomic diagrams of Abdel-Malek (1955) and Malek (1985) will prove useful (Fig. 5–5).

For schistosomes, and probably for other trematodes also, the range of snail hosts suitable for the development of any particular geographic isolate is far more restricted than that of vertebrate hosts. Whereas *S. japonicum,* for example, can infect mice, rats, rabbits, dogs, pigs, cattle, and various wild mammals in addition to humans, it is restricted to limited subspecies of *Oncomelania hupensis* for its intramolluscan development. Isolates of *S. mansoni* are even more limited in intermediate host range, with varying compatibility between particular populations of parasites and of *Biomphalaria*

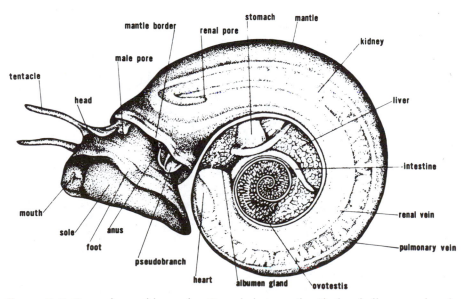

FIGURE 5–5 General morphlogy of a *Biomphalaria* snail with the shell removed and viewed from the left side. *Source:* Malek 1985.

snails (Basch 1976). For *S. haematobium* obtained from different regions of the Middle East and Africa,

> certain species or populations of snails from the genus *Bulinus* are incapable of acting as intermediate hosts for schistosomes from the same or other geographical regions, whereas some snails show a high degree of compatibility with other strains of the parasite transmitted elsewhere by snails belonging to the same species complex or subspecies. (Paperna 1967)

Miracidial Penetration and Conversion to the Primary Sporocyst

Observation of known compatible miracidia-snail interactions through the dissecting microscope is often frustrating because many miracidia are seen to come in contact with the foot, face, tentacles, pseudobranch, or other exposed parts of the snail, to probe or wander around for a short time, and then swim away. Some of these may return to penetrate successfully, while others appear foolishly to have thrown away their only chance at survival.

A miracidium attaches to the snail's skin by its apical papilla (= terebratorium, rostrum, anterior papilla) (Fig. 5–6). Kinoti (1971) considered the apical papilla as an attachment organ "functionally analogous to the suckers of the adult schistosome. Penetration into the body is probably achieved by a combination of mechanical movement and secretions of the unicellular structures variously called penetration, adhesive or cephalic glands." The functional relationship, if any, between these glands and the so-called apical gland remains to be clarified.

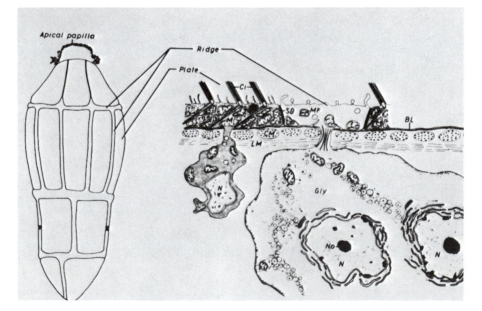

FIGURE 5–6 Outer structure (left) and body wall ultrastructure (right) of free miracidium of *S. mansoni*. BL, basement membrane; Ci, cilia; CM, circular muscle; Gly, glycogen; Mi, mitochondrion; N, nucleus; No, nucleolus; SD, septate desmosome. *Source:* Meuleman et al. 1978.

After penetration that ciliated plates (Fig. 5–6) are thrown off and a new surface tegument, that of the sporocyst, is produced from material mobilized from subepithelial cells (Basch and DiConza 1974). The new surface is soon covered with microvilli, which become abundant and branched by 20 hours of incubation in vitro. Presumably this surface carries the particular stereo-chemical configurations that determine whether or not the newly trans-formed sporocyst will survive or perish within its snail host.

Permissive and Nonpermissive Snail Hosts

A worldwide biological matrix is established by an infinity of local variants of trematodes and "potential" host snails. For *S. mansoni* Basch (1976) re-viewed the extensive earlier literature, including numerous laboratory trials in which snails from one geographic area were exposed to miracidia from the same or other areas, with resultant infection rates ranging from nil to almost 100%.

Work on schistosome-snail compatibility in the early decades of the twentieth century was reviewed by Cram et al. (1947). These authors also exposed 27 species and subspecies of native American snails to *S. mansoni,* finding only a species identified as *Tropicorbis havanensis* susceptible, in addition to the usual host, *Australorbis glabratus* (both are now in *Biomphalaria*). Shortly thereafter, Files and Cram (1949) and Files (1951) exposed *Biomphalaria* from the Caribbean, South America, and Africa to *S. mansoni* of Puerto Rican, Venezuelan, Egyptian, and other origin, finding great dif-ferences in infection rates.

A histological study was undertaken by Brooks (1953), who exposed *T. havanensis* from Louisiana and *B. glabrata* from Puerto Rico to a Puerto Rican isolate of *S. mansoni*. Miracidia penetrated equally well and larvae invaded the tissues of all exposed areas of both snails, but

> Host-tissue response produced encapsulation of a great majority of the miracidia or sporocysts in *T. havanensis*. By examining snails sacrificed at appropriate intervals it was found that most of the parasites were destroyed between three and fifteen days after their entry into the host. Only a few survived beyond seven days. . . . Tissue response, regardless of the number of invading larvae, was not seen in *A. glabratus*. . . . it is suggested that the occasional successful develop-ment of sporocysts in [*T. havanensis*] may be due more to intrinsic variation in the miracidia than to their chance location in exceptionally susceptible individual snails or in particularly favorable host tissues. (Brooks 1953)

Simultaneously, Newton (1952) carried out a similar study but with a susceptible Puerto Rican *B. glabrata* and refractory Brazilian snails of the same species, finding

> In the susceptible Puerto Rican strain, the parasite develops apparently without any effective interference on the part of the snail. In contrast, with the Brazilian strain the parasite is destroyed and removed, usually within 24 to 48 hours after penetration. As at least part of this response, there is a marked cellular infiltra-tion around the parasite. A fibrotic type of walling-off also appears to be char-acteristic of this response. (Newton 1952)

In the same era DeWitt (1954) studied compatibilities among *S. japoni-cum* from China, Japan, Taiwan, and the Philippines, and potential inter-mediate host snails from the same areas plus North America, with expected variations in infection rates.

It is likely that all larval trematodes exhibit similar relationships with their own molluscan intermediate hosts and with related forms that, to hu-mans, may appear indistinguishable. Studies over the past 4 decades, while supplying suggestive evidence, have fallen short of confirming the specific mechanisms of compatibility or resistance.

The Biological Basis of Schistosome-Snail Compatibility

Frandsen (1979), finding a disparity of methods and materials in use in var-ious laboratories, made a plea for standardization of 1) experimental mate-rials such as schistosomes, intermediate and definitive hosts; 2) methods involved in the experiments; and 3) data collection. There is little evidence that Frandsen's appeal has met with much collaboration.

A short list of possible explanations for schistosome-snail compatibility was proposed by Bayne and Yoshino (1989):

TABLE 5–1 Phenomena that May Explain Compatibility in Molluscan Schistosomiasis

1. Sporocysts resist toxic components of the host
2. Sporocysts avoid being recognized by the host due to
   a. Molecular mimicry
   b. Acquisition of host molecules
   c. Prevention of opsonization
3. Sporocyts interfere with host hemocyte function

There is variably convincing evidence for and against each of these ideas. The current consensus views snail-parasite compatibility as a manifestation of invertebrate immunology (van der Knaap and Loker 1990), in which host hemocytes (amebocytes) attack and destroy invading sporocysts recognized as foreign (Bayne 1982), probably in concert with soluble factors in the mol-luscan hemolymph (Bayne 1980). This process has been demonstrated to occur in a "partial in vitro" system in which flat glass capillary tubes con-taining compatible and incompatible miracidia were implanted as indwelling cannulae into the hemolymph sinus of living *Biomphalaria* snails (Fig. 5–7) (Basch 1979). The concept has been modified more recently:

An emerging hypothesis evolving from work on larval E-S [excretory-secretory] products is that differences in the innate susceptibility of snail strains to a specific trematode stain may be a reflection of how hemocytes perceive and respond to a specific molecular signal from the parasite. . . . perhaps both stimulatory and inhibitory factors are secreted simultaneously by invading larvae, and the "net" behavioral response of circulating hemocytes to the parasite may depend upon their relative sensitivity to each factor. (Bayne and Yoshino 1989)

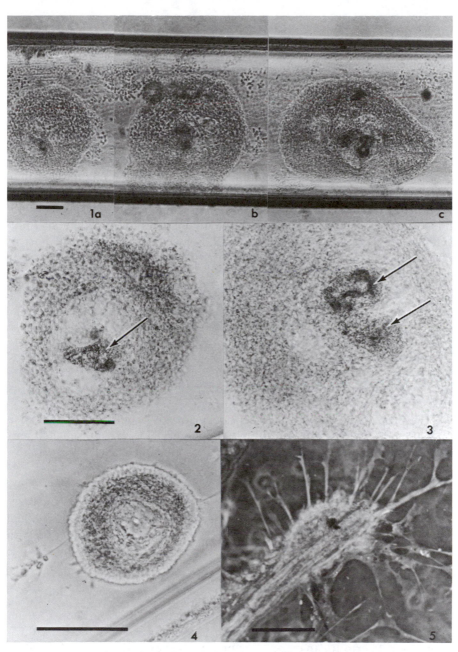

Figure 5–7 Granulomas formed around sporocysts of *S. mansoni* in a capillary tube inserted into the hemolymph sinus of *B. glabrata* snails. Phase contrast photographs of living cells. 1a, b, c, Three granulomas from a resistant R1-R4 snail. Scale bar = 100 μm; 2, 3 Details of 1b and 1c, respectively. Arrows indicate the remains of sporocysts. Scale bar = 100 μm. 4, Granuloma in glass capillary tube removed from a resistant 10-R2 snail showing the lamellar structure and sharp periphery. 5, Cells forming around a foreign object (fiber) in a tube removed from a susceptible PR-B snail. *Source:* Basch 1979.

It is likely that populations of both schistosomes and their snail hosts are polymorphic at many genetic loci, and that compatible (i.e., permissive) combinations are based on genetically determined phenotypic concordances. Any of the specific mechanism(s) of rejection mentioned above, or others, could be operative under such a system. Basch (1975) has considered these interactions in a generalized mathematical formulation:

Disregarding environmental conditions, each encounter between a potential host snail and a miracidium will result in success (S) leading to normal development if both are of compatible phenotypes, and in failure (F) if they are not. Infection (I) is defined as one or more successes in a series of N separate encounters. A defined population of host snails may present with phenotypes H_1, H_2, . . . H_n, and a defined population of miracidia may present with phenotypes, M_1, M_2, . . . M_n. For each individual snail host of phenotype H_i there may be anything from no to all compatible phenotypes among the miracidial population.

Let x_j = the proportion of miracidia of phenotype M_j.

Let $y_i = \Sigma_j x_j$ for the M's that are compatible with H_j.

The probability of success in a snail of phenotype H_i that encounters one miracidium from the parasite population is given as $P(S_i) = y_i$.

Therefore, $P(F_i) = (1 - y_i)$; and the probability of infection $p(I_i) = 1 - (1 - y_i)$.

The probability that infection will occur in any randomly selected snail from the defined host population that encounters N randomly selected miracidia from the defined parasite population is given by Basch (1975) as

$$P(I) = \sum_{i=1}^{n} P(H_i) \times 1 - (1 - y_i)^N.$$

These formulas hold for any snail-miracidium combination. Individual snails, colonies, or populations may be highly susceptible to infection with one trematode while resistant to another, in relationships that can change over time.

Genetics of Compatibility. The pioneering work of Newton (1953) established standard laboratory conditions under which 95% of a Puerto Rican, and zero % of a Brazilian isolate of *B. glabrata* became infected with *S. mansoni* of Puerto Rican origin. Reciprocal crosses of these hermaphroditic snails (guarding against self-fertilization) showed varying infection rates until F_3 populations were derived with infection rates ranging from 0% to 82%. Newton demonstrated for the first time that susceptibility to *S. mansoni* in *B. glabrata* "is a heritable character, and that several genetic factors are probably involved." More recently, the genetics of *S. mansoni-B. glabrata* schistosome-snail compatibility have been studied by Richards (e.g., 1975, 1984), Mulvey and Woodruff (1985), and others.

In some *B. glabrata-S. mansoni* combinations the snail is actively resistant. It recognizes the miracidium as foreign, and amebocytes rapidly encapsulate and destroy the parasite. In some snail-parasite combinations the miracidium evades recognition and encounters little or no host response. In the absence of active resistance, if the host hemolymph is unsuitable for normal development, the parasite may survive briefly then die. In some snail-parasite combinations, parasite development is temporarily delayed but is eventually resumed with production of cercariae. In a nonresistant and suitable snail host the parasite develops normally. Snail crosses suggest that active resistance is dominant over lack of resistance, and in the absence of host response, that unsuitability is dominant over suitability. (Richards and Shade 1987)

Four susceptibility types have been distinguished: I, nonsusceptible at any age; II, juvenile susceptible/adult nonsusceptible; III, susceptible at any age; IV, juvenile susceptible, adult variable (Richards 1977, 1984). Age-dependent susceptibility changes occur also in other snail-schistosome combinations. For example, juveniles of European *Lymnaea stagnalis* are highly susceptible to infection with *Trichobilharzia ocellata,* but adults become refractory (Meuleman and te Velde 1984). By inbreeding over long periods, Richards was able to develop snail stocks of either zero or close to 100% infectivity with the PR-1 (Puerto Rican origin) strain of *S. mansoni.* Most inbred snail stocks exhibited differing infection rates when challenged with miracidia of other isolates of *S. mansoni* from Puerto Rico, or from Egypt, Kenya, or Zaire. One was nonsusceptible to all *S. mansoni* strains, and one was susceptible to all. Several genetic loci in each organism, some involving multiple alleles, are involved in the control of host-parasite compatibility (Richards and Shade 1987).

The crucial parameter to the parasite is the number of cercariae shed by any snail invaded by a miracidium, as a factor determining the likelihood of success in completing the life cycle. Ward et al. (1988) looked at total cercarial production in two populations of *B. glabrata,* one 90–100% susceptible and the other 70–80% susceptible. More miracidia established successful primary sporocyst infections in the former group, which released nearly twice the number of cercariae per infected snail. However, snails of both groups infected with single miracidia produced equal numbers of cercariae, suggesting that all primary sporocysts have the same reproductive potential.

The influence of snail size, as opposed to age, on infection was studied by Niemann and Lewis (1990). Decreased susceptibility was established in larger, not in older, snails. Total cercarial production per infected snail was uniform, perhaps because more sporocysts developed in smaller snails but more host tissue per sporocyst is available for exploitation in larger ones.

Mechanism of sporocyst destruction

Numerous authors (e.g., Bayne et al. 1987) have found widespread antigenic cross-reactivity between protein components of schistosomes and their intermediate host snails, suggestive of antigenic mimicry (Yoshino and Bayne

1983) but this finding alone does not provide evidence for any specific immune-recognition system. Similarly, the finding of putative antigen binding receptor sites on snail hemocytes (Bayne et al. 1984) is suggestive, but not confirmatory, of the recognition hypothesis.

In a comprehensive review, van der Knaap and Loker (1990) discussed four distinct categories of host cells that play a role in the internal defense system of snails. Three of these (antigen-trapping cells, reticulum cells, and pore cells) are fixed in the tissues and their role vis-à-vis larval trematodes is unclear. The fourth category, which includes mobile hemocytes of several subtypes, is specifically involved in snail-trematode interactions.

Molluscan hemocytes are produced in various hemopoietic tissues (Sminia 1981). In *B. glabrata* an amebocyte-producing organ has been described by Lie et al. (1975) between the pericardium and posterior epithelium of the mantle cavity. The resting organ is small, thin, and inconspicuous, but marked hypertrophy and hyperplasia occur soon after infection or reinfection of the snail. The ultrastructure of this organ was described by Jeong et al. (1983). Small clusters of primary ameboblasts are held in a loose reticulum in the lining of the pericardium. On activation, a mitotic burst occurs to produce large clusters of primary and secondary ameboblasts. Fully activated amebocyte producing organs consist of masses of progressively maturing cells, significantly larger than those found in resting organs.

The role of the amebocyte-producing organ and host amebocytes has been explored by whole-body irradiation in order to increase the susceptibility of otherwise refractory snails. Michelson and DuBois (1981) could not raise infection rates with *S. mansoni* in their Brazilian *B. glabrata*; indeed, irradiation actually reduced infections in a susceptible population of snails. In contrast, Sullivan and Richards (1982) were able to show a somewhat enhanced infection rate with the trematode *Ribeiroia marini* after irradiation of *B. glabrata*. These interactions have been reviewed by Bayne (1983).

Direct hemocyte-sporocyst interaction, observed at the light microscope level in the 1950s by Brooks, Newton and others, has been studied ultrastructurally in various snail-schistosome systems. Krupa et al. (1977), looking at *S. haematobium* in *Bulinus guernei* found two types of hemocytes: granulocytes and hyalinocytes, associated with the sporocysts, but only granulocytes interacted with microvilli on the surfaces of sporocysts. With a *B. glabrata-S. mansoni* system Loker et al. (1982) found hemocyte contact within 3 hours and phagocytosis of sporocyst microvilli and pieces of tegument within 7.5 hours of infection. Extensive pathology was demonstrated within 24 hours and by 48 hours only scattered remnants of sporocysts remained. Development of amebocytes and amebocytic accumulations are presumably under genetic control (Richards 1980).

The review of van der Knaap and Loker (1990) summarizes current knowledge of hemocyte function. Recognition of foreignness in snails may be mediated through lectins produced by hemocytes and released into the plasma, but specific mechanisms remain to be clarified. Accumulations of hemocytes around incompatible sporocysts suggests a chemotactic capability, and recruitment of additional cells produces a multilatered capsule that may be benign or lethal to the sporocyst, depending on hemocyte activation to a cytotoxic state. Actual killing of sporocysts recognized as foreign may be mediated by peroxidases or other enzymes released by the hemocytes, and/or by physical attack. Possible evasion mechanisms of sporocysts are listed on Table 5–1.

Development of Sporocysts

Primary Sporocysts. Whitfield and Evans (1983) have considered multiplication in digenean germinal sacs to be basically an asexual budding process, as is the production of cercariae within the daughter sporocysts. Intramolluscan stages were interpreted by Clark (1974) similarly as a type of budding, rather than as an alternation of generations or polyembryony. By maintaining a relatively small number of totipotent cells, which divide infrequently, the intramolluscan stages are viewed as being spared to an extent from the incorporation of genetic errors.

Although fasciolids and some other trematode miracidia shed their ciliated plates in the surrounding water during penetration, the entire schistosome miracidium enters the snail. The ciliated miracidial plates are cast off within 2 hours in *S. mansoni,* and presumably also in other species of schistosomes. Once within the snail tissue, the miracidium is transformed into the primary (or mother) sporocyst. Narrow ridges in the areas between the former ciliated plates expand as subtegumentary material rises and spreads over the surface to form the continuous outer layer of the sporocyst (Basch and DiConza 1974; Meuleman et al. 1978). The precise biochemical configuration of the interface presented to the host appears to be critical for continued survival in that individual snail.

Miracidia of *S. mansoni* placed in phosphate-buffered saline or in a variety of cell culture media begin to transform to sporocysts by casting off their ciliated plates within 3 hours at 22 C, but do so only after at least 2 minutes of prior exposure to (artificial) pond water (Samuelson et al. 1984). Schutte (1974b) reported that the terebratorium, musculature, sensory papillae, acidophilic cells and excretory ampulla also disappear within the first 24 hours. The neural mass, apical and penetration glands disappear over the following days in a variable pattern. As described above, miracidial subepidermal cells give rise to the syncytial tegument of the primary sporocyst.

The development of the primary sporocyst and secondary (or daughter) sporocysts of *S. mansoni* has been described by Schutte (1974b) by light microscopy (Fig. 5–8), and on an ultrastructural level by Meuleman et al. (1980) and by Göbel and Pan (1985). In vitro models of these processes few (e.g., Di Conza and Basch 1974, Basch and DiConza 1974; Mellink and van den Bovenkamp 1985), and further experimental studies are needed.

Few studies have been done on the physiology of miracidial transformation mechanisms. Using *S. mansoni* miracidia transformed by saline, ammonium salts, or imidazole, Kawamoto et al. (1989) found that the initiation of this process may be synergistically regulated by cyclic AMP and Calcium^{++} ions.

A method of labeling miracidia with tritiated leucine was developed by Kassim and Richards (1979) that permitted the primary sporocysts to be found readily by autoradiography of the snail tissues. Tritiated thymidine was not taken up, indicating a lack of synthesis of DNA by miracidia. In most schistosome species studied (e.g., *S. mansoni, S. haematobium, S. intercalatum, S. bovis*), primary sporocysts usually develop near the point of miracidial entry, particularly in exposed tissues of the foot margin, face, mantle edge, and tentacles (Fig. 5–5). In the case of *S. japonicum*, they may be found almost anywhere in the host *Oncomelania*, with a preference for

---

FIGURE 5–8 Development of the sporocysts of *S. mansoni*. 1, Young primary sporocysts with persisting miracidial components. a, b, 24 hours old; c, d, 48 hours old. 2, Primary sporocysts a, 3 days old in rectal sinus; b, 6 days old in foot; c, 7 days old in foot, containing embryonic secondary sporocyst. 3, Advanced primary sporocysts a, 14 days old in foot, containing embryos in various stages of development; b, same, extremity of sporocyst; c, 16 days in foot, showing size of reticulum cells. 4, a, mature secondary sporocyst within primary sporocyst, 18 days old; b, migratory secondary sporocyst 20 days old in peri-rectal sinus, with attached host amebocytes and fibroblasts; c, more compact migratory secondary sporocyst 20 days old in buccal mass; d, portion of primary sporocyst in foot with two mature secondary sporocysts; e, advanced secondary sporocyst 20 days old, apparently trapped in kidney. Scale bar = 50 μm. 5, a, Portion of migratory secondary sporocyst 20 days old in columella muscle, with strings of granular material and reticulum and small ganular cells attached to the wall by cytoplasmic connections; b, secondary sporocyst 18 days old immediately after arrival in interlobular space of the digestive gland; c, portion of advanced sporocyst 27 days old in digestive gland, showing a flame cell; 6, Advanced secondary sporocysts in interlobular spaces of the digestive gland a, 18 days old, with several early cercarial embryos; b, 18 days old, with more advanced cercarial embryos; c, 27 days old retaining a few uncleaved germinal cells in the parietal layer. AC acetabulum, AG apical gland, AM amebocyte, AN anterior gland, BG basophilic granules, BW body wall, C cilia Ca cecum, CEM embryonic cercariae, CI nucleus with cytoplasmic connection, CM circular muscle, CR crura of primary sporocyst, DL digestive lobule, EG escape gland, ES embryonic secondary sporocyst, ET excretory tubule, F foot of snail, FB fibroblast, FC flame cell, FR furcal rami, GA primitive gut anlage, GM granular material and reticulum fibers, GN granular nucleus, GPC primpridial gland cell, GR granular material, ID tegumental nuclei, IN indentation, IS intensely staining unit, IV investing cell, KE kidney eoithelial cell, LA lacunae, LM longitudinal muscle, M membrane, MA matrix, MC mantle collar, ME mesomere, MI micromere, MIT mitotic cell or nucleus, MM macromere, MS metasporocyst, MU muscle layer, NG ganglia of snail, NM neural mass, NU nucleus, OA ovotestis acinus, OS oral sucker, PA parenchymal cell, PAR parietal layer, PG penetration gland, PM primordial muscle cell, PN peripheral nuclei, PO posterior gland, PR primitive gut, PS primary sporocyst, RET reticulum cell, RF reticulum fibers, SP spines, SS secondary sporocyst, SR sphincter, T tail stem, UD unequal division, V vacuole. *Source:* Schutte 1974b.

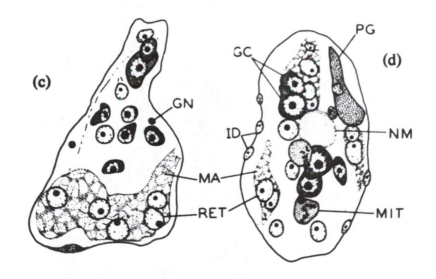

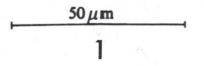

1

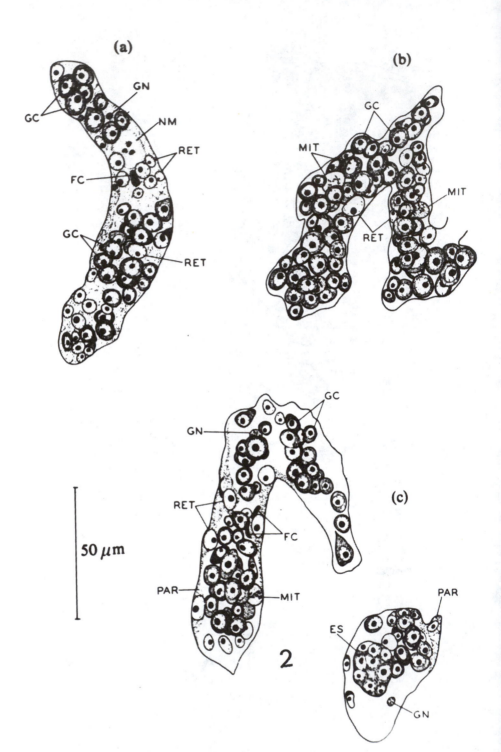

FIGURE 5–8 (continued)

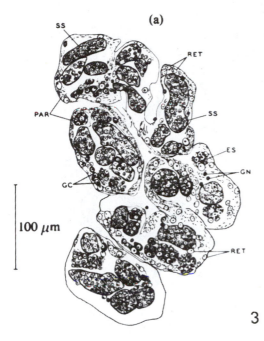

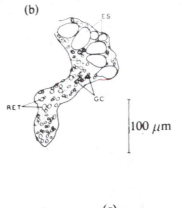

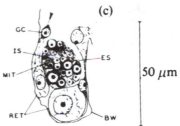

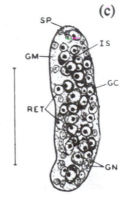

3

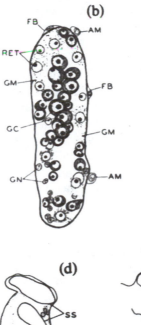

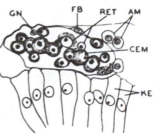

4

FIGURE 5–8 *(continued)*

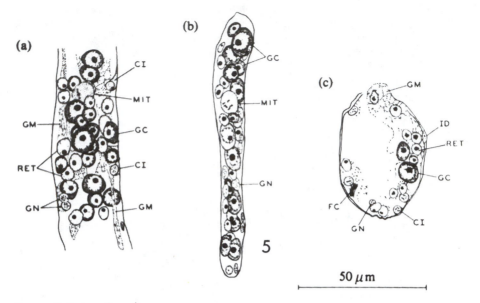

FIGURE 5–8 (continued)

loosely organized tissues and hollow organs, including the branchial, renal, and circulatory systems and female genital tract (but not the glands) (Jourdane and Xia 1987). There is no barrier to penetration by more than one miracidium, as demonstrated by the possibility of bisexual infections in animals exposed to cercariae from a single snail exposed to several miracidia in the field or laboratory. Despite intentional exposures to hundreds of miracidia, infections with more than a handful of primary sporocysts are probably rare in the *Biomphalaria glabrata-Schistosoma mansoni* combination. In this situation it is possible, but improbable, that heavily infected snails die quickly and escape observation; it seems more likely that some control system is operative to limit the number of entering miracidia that develop to primary sporocysts. Jourdane and Xia (1987) postulated that no such system exists in the *O. h. hupensis-S. japonicum* system, as they found one snail with 32 primary sporocysts.

Secondary Sporocysts. A detailed and extensively documented ultrastructural study of development of *S. japonicum* sporocysts in *Oncomelania* snails was carried out by Göbel and Pan (1985) (Fig. 5–9). An analogous electron microscope study of *S. mansoni* sporocysts in *B. pfeifferi* was done by Meuleman (1972). The intensive light-level investigation of the same species by Schutte (1974b) was beautifully illustrated with line drawings of sporocysts (Fig. 5–8). Schutte (1974b) divided sporocyst development into two phases: pre-hepatic with (a) increase in germinal cells; (b) first embryos appear in primary sporocysts; (c) embryos begin to elongate; (d) primary sporocysts mature; (e) secondary sporocysts emerge from primary sporocysts,

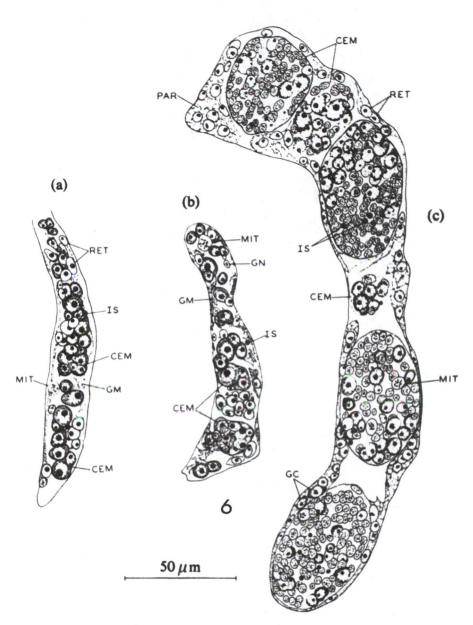

(a)

(b)

(c)

6

50 μm

FIGURE 5–8 (continued)

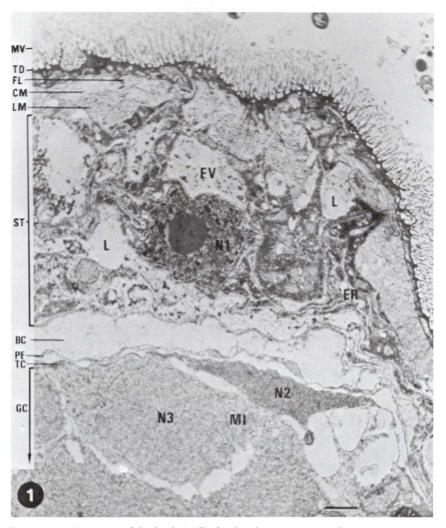

Figure 5–9 Structure of the body wall of a daughter sporocyst of *S. japonicum* containing a germ ball, 2 months post infection. BC, brood chamber; CM, circular muscle fiber; ER, endoplasmic reticulum; FL, fiber layer; FV, food vacuole; GC, germinal cell; L, lipid vacuole; LM, longitunal muscle fiber; MI, mitochondrion; MV, microvilli; NI, nucleus of subtegumental cell of daughter sporocyst; N2, nucleus of tegument of cercarial embryo; N3, nucleus of germinal cell; PE, primitive epithelium of cercarial embryo; ST, subtegument; TC, tegument of cercaria; TD, tegument of daughter sporocyst. Scale bar = 1 μm. *Source:* Göbel and Pan 1985.

and hepatic with (f) secondary sporocysts arrive in digestive gland; (g) embryos appear in metasporocysts; (h) cercarial differentiation; (i) morphologically mature cercariae; and (j) cercarial shedding. Each stage is described in detail. The term "metasporocyst" has not been widely adopted.

Escape of cercariae of *S. mansoni* through a birth pore in the daughter sporocyst has been documented and illustrated by Fournier and Théron (1985) Fig. 5–10.

Number of Secondary Sporocysts. Upatham (1973a) carried out an experiment in which he exposed first-generation laboratory-bred *B. glabrata* snails individually to 1, 2, 4, 8, 16, or 32 miracidia of *S. mansoni* (all from St.

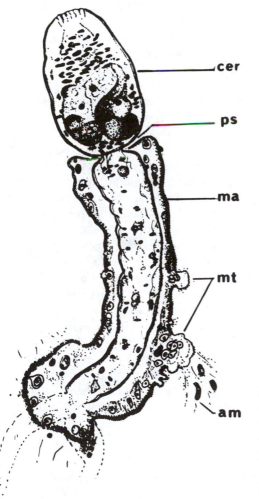

FIGURE 5–10 Passage of a cercaria through the anterior thickened area of the daughter sporocyst of *S. mansoni.* am, amebocytes; cer, cercaria; ma, anterior thickened area; mt, tegumentary microlesion; ps, birth pore. *Source:* Fournier and Théron 1985.

Lucia) for 18 hours, and counted the number of resultant secondary or daughter sporocysts 12 days later:

Miracidia per Snail	Mean and S. E.		Range
1	22	(6.06)	9–34
2	74	(14.14)	40–98
4	128	(8.69)	110–158
8	137	(12.40)	112–176
16	220	(51.62)	162–520
32	359	(62.85)	210–620

Multiplication of Secondary Sporocysts. It was formerly believed that a single generation of secondary sporocysts intervened between the primary sporocyst and cercariae, but the work of Hansen (1973) and Hansen et al. (1973) established the existence of "progeny-daughter sporocysts" in *S. mansoni.* Techniques for transplantation of sporocysts of this species from infected to uninfected *B. glabrata* were reported by Chernin (1966) and DiConza and Hansen (1972), and more recent work of this nature has demonstrated the great capacity for multiplication of the secondary sporocysts. Successful implantations have been made of *S. mansoni* (Jourdane 1982, 1983, 1984; Jourdane et al. 1980, Jourdane and Théron 1980, Imbert-Establet et al. 1984, Cohen and Eveland 1984, 1988), *S. haematobium* (Jourdane et al. 1981, Kechemir and Combes 1982), *S. bovis* (Jourdane et al. 1984), and *S. japonicum* (Jourdane et al. 1985); as many as 7 consecutive passages have been reported. Recipient snails eventually come to resemble those normally infected, in that the digestive gland ovotestis, and other tissues become filled with secondary sporocyst material. Two mechanisms may be responsible for this proliferation: growth, budding, or fragmentation of preexisting sporocysts, and development of true subsequent sporocyst generations within the transplanted larvae. Both processes may occur. These experiments demonstrate that the volume of secondary sporocysts eventually derived from a single miracidium is determined by the size of the snail host and not by any inherent limit in generative capacity of the primary sporocyst.

It has been shown by histological studies that additional generations of sporocysts form within normal secondary sporocysts in both *S. mansoni* (Théron and Jourdane 1979) and *S. haematobium* (Kechemir and Théron 1980). Sporocyst transplantation will be useful in genetic studies on schistosomes because particular clones originating in unimiracidial infections may be maintained long beyond the normal life span of any single infected snail.

Interactions Among Larval Trematodes Within the Same Snail

It has been pointed out that trematode miracidia of any species and genetic isolate can develop in only a very restricted range of snail hosts. Conversely, each individual susceptible snail may be penetrated and infected in any order

by one or more miracidia of those species for which it can serve as intermediate host. K.J. Lie and Paul F. Basch, working in Malaysia on the life
cycles of different trematodes (*Echinostoma audyi* and *Trichobilharzia
brevis*) transmitted in the same habitats by *Lymnaea rubiginosa*, exposed
those snails in varying patterns to laboratory-reared miracidia of the two
species. They reported (Basch and Lie 1965; Lie et al. 1965; Basch and Lie
1966a, b) the occurrence of antagonistic interactions within the tissues of *L.
rubiginosa* that resulted regularly in the disappearance of one parasite (*T.
brevis*) and the predominance of the other (*E. audyi*). The mechanism of this
interaction was evident. Rediae of echinostomes, actively motile and possessing a well-developed mouth and muscular pharynx, were observed lit-
erally to prey on the sessile and essentially defenseless schistosome sporocysts until the latter had all been attacked and consumed. The same fate
awaited strigeid and plagiorchid larvae in snails coinfected with the echinostome. This process of hunting down and eating competing larval trematodes
was termed direct antagonism.

Field collections in Malaysia had shown that there were four species of
echinostomes found within *L. rubiginosa* in local aquatic habitats, but no
individual snail was ever found naturally infected with more than one of
them. Subsequently, additional double infections were established with any
of three echinostomes—*Echinostoma barbosai, E. lindoense,* or *E. paraensei*—together with the predatory rediae of the echinostome *Paryphostomum
segregatum,* in which the *P. segregatum* was found regularly to be dominant
(Lie et al. 1968a).

Basch et al. (1969) established double infections in *Biomphalaria glabrata* with two species, *Schistosoma mansoni* and the strigeid *Cotylurus
lutzi,* in which only sporocysts, and no rediae, are found. The *S. mansoni*
exerted a strong indirect antagonism causing degeneration of the strigeid
sporocysts. The mechanism of indirect antagonism is still unknown but
might involve competition for nutrients, production of specifically inhibitory
materials, or mediation through host humoral or cellular responses. A review of these and other earlier studies was published by Lim and Heyneman
(1972).

Inter-trematode antagonism may occur in any habitat in which several
parasite species cycle through the same snail hosts. Hata et al. (1988) infected *Oncomelania nosophora* snails with *S. japonicum* and *Paragonimus
ohirai,* with findings typical of double-infection experiments. When snails
already infected with *P. ohirai* were exposed to *S. japonicum* miracidia, resultant infection rates were low (0–7%), sporocysts were few, degenerate,
and transient, and no cercariae were produced, in contrast with 55 to 75%
normal infections in *S. japonicum*-exposed controls. Simultaneous double
exposures resulted in infection with both species but the subsequent development of *S. japonicum* was less than in controls. However, when an *S.
japonicum* infection preceded exposure to *P. ohirai* miracidia, both species
developed to cercariae as the young rediae of *Paragonimus* were apparently
unable to overcome the established *S. japonicum* infection.

As a rule snails infected with a dominant parasite cannot be reinfected with a subordinate species, while those already harboring a subordinate species can be reinfected readily with a dominant one (Lie 1973). Presuming that the schistosomes are generally subordinate, this phenomenon suggests a means of biological control in which eggs of an "innocuous" dominant trematode of nonveterinary and nonhuman significance could be intentionally distributed in an area to reduce or eliminate snail infections with economically or medically important schistosomes (Lie et al. 1968c; Lim and Heyneman 1972). In interactions between *S. mansoni* and the bird parasite *Ribeiroia marini* (family Cathaemasiidae), sympatric with *S. mansoni* in areas of the Caribbean, *R. marini* was partially dominant over *S. mansoni* (Page and Huizinga 1976). All snails with double infections shed significantly fewer schistosome cercariae and had fewer sporocysts than singly infected controls, but the schistosome was not eliminated. The rediae of *R. marini* in doubly infected snails were retarded in development and had distorted teguments, showing an indirect antagonism in the opposite direction. It remains to be seen whether this phenomenon could help to provide practical schistosome control in field situations, but experiments of this type may help to explain the spotty distribution of schistosomes within larger regions occupied by otherwise suitable host snails.

References

Abdel-Malek ET. 1955. Anatomy of Biomphalaria boissyi as related to its infection with Schistosoma mansoni. American Midland Naturalist 54:394–404.

Abdel-Salam E, Abdel Khalik A, Abdel-Meguid A, Barakat W, Mahmoud AA. 1986. Association of HLA class I antigens (A1, B5, B8 and CW2) with disease manifestations and infection in human schistosomiasis mansoni in Egypt. Tissue Antigens 27:142–146.

Abou Senna HO, Bassaly M. 1964. [Observations on the hatching of miriacidia from schistosome eggs.] Medizinskaia Parazitologii 33:704–706.

Adam W. 1960. Contribution à la connaissance du développement sexuel et de l'hermaphroditisme chez les schistosomes. Institut Royale des Science Naturelles de Belgique, Bulletin 36 No. 31:1–22.

Alves W. 1947. Observations on S. mattheei and S. haematobium. Adults and eggs from experimental animals and man. Transactions of the Royal Society of Tropical Medicine and Hygiene 41:430–431.

Alves W. 1949. The eggs of Schistosoma bovis, S. mattheei and S. haematobium. Journal of Helminthology 23:127–134.

Amin MA, Nelson GS. 1969. Studies on heterologous immunity in schistosomiasis. 3. Further observations on heterologous immunity in mice. Bulletin of the World Health Organization 41:225–232.

Anderson LA, Cheever AW. 1972. Comparison of geographical strains of Schistosoma mansoni in the mouse. Bulletin of the World Health Organization 46:233–242.

Anderson MG, Anderson FM. 1963. Life history of Proterometra dickermani Anderson, 1962. Journal of Parasitology 49:275–280.

Appleton CC. 1982. The eggs of some blood-flukes (Trematoda: Schistosomatidae) from South African birds. South African Journal of Zoology 17:147–150.

Armstrong JC. 1965. Mating behavior and development of schistosomes in the mouse. Journal of Parasitology 51:605–615.

Aronstein WS, Strand M. 1984. Gender-specific and pair-dependent glycoprotein antigens of *Schistosoma mansoni*. Journal of Parasitology 70:545–557.

Aronstein WS, Strand M. 1985. A glycoprotein antigen of *Schistosoma mansoni* expressed on the gynecophoral canal of mature male worms. American Journal of Tropical Medicine and Hygiene 34:508–512.

Asch HL, Dresden MH. 1979. Acidic thiol proteinase activity of *Schistosoma mansoni* eggs. Journal of Parasitology 65:543–549.

Asch HL, Read CP. 1975a. Transtegumental absorption of amino acids by male Schistosoma mansoni. Journal of Parasitology 61:378–379.

Asch HL, Read CP. 1975b. Membrane transport in Schistosoma mansoni: Transport of amino acids by adult males. Experimental Parasitology 38:123–135.

Atkinson BG, Atkinson KH. 1982. *Schistosoma mansoni*: One and two-dimensional electrophoresis of proteins synthesized in vitro by males, females, and juveniles. Experimental Parasitology 53:26–38.

Atkinson KH, Atkinson BG. 1980a. Sex differences in polypeptides synthesized in vitro by *Schistosoma mansoni*: One- and two-dimensional electrophoresis. Canadian Journal of Genetics and Cytology 22:656.

Atkinson KH, Atkinson BG. 1980b. Biochemical basis for the continuous copulation of female *Schistosoma mansoni*. Nature 283:478–479.

Awad AH, Probert AJ. 1989. Transmission and scanning electron microscopy of the male reproductive system of Schistosoma margrebowiei Le Roux. Journal of Helminthology 63:197–205.

Awwad M, Bell R. 1978. Faecal extract attracts copulating schistosomes. Annals of Tropical Medicine and Parasitology 72:389–390.

Azimov DA. 1970. (Revision of the system of trematodes of Schistosomatata Skrjabin et Shulz, 1937). Zoologicheskii Zhurnal 49:1126–1130.

Bair RD, Etges FJ. 1973. *Schistosoma mansoni*: factors affecting hatching of eggs. Experimental Parasitology 33:155–167.

Bang FB, Hairston NG. 1946. Studies on schistosomiasis japonica IV. Chemotherapy of experimental schistosomiasis japonica. American Journal of Hygiene 44:348–366.

Barrabes A, Duong TH, Combescot C. 1979. Effet de l'administration d'implants de testosterone ou de progesterone sur l'intensité de la parasitose experimentale à *Schistosoma mansoni* du hamster doré femelle. Comptes Rendus des Séances de la Société de Biologie et des ses Filiales (Paris) 173:153–156.

Barshene IaV, Staniavichiute GI, Orlovskaia OM. 1989. [Karyologic research on trematodes of the family Schistosomatidae from northwestern Chukota]. Parazitologiia 23:496–503. In Russian.

Basch PF. 1966. The life cycle of *Trichobilharzia brevis*, n. sp., an avian schistosome from Malaya. Zeitschrift fur Parasitenkunde 27:242–251.

Basch PF. 1975. An interpretation of snail-trematode infection rates: specificity based upon concordance of compatible phenotypes. International Journal for Parasitology 5:449–452.

Basch PF. 1976. Intermediate host specificity in *Schistosoma mansoni*. Experimental Parasitology 30:150–169.

Basch PF. 1979. Biomphalaria hemocyte migration and parasite encapsulation in implanted flat glass tubes. Journal of Invertebrate Pathology 34:99–101.

Basch PF. 1981a. Cultivation of *Schistosoma mansoni* in vitro. 1. Establishment of cultures from cercariae and development until pairing. Journal of Parasitology 67:179–185.

Basch PF. 1981b. Cultivation of *Schistosoma mansoni* in vitro. 2. Production of infertile eggs by worm pairs cultured from cercariae. Journal of Parasitology 67:186–190.

Basch PF. 1984. Development and behavior of cultured *Schistosoma mansoni* fed on human erythrocyte ghosts. American Journal of Tropical Medicine and Hygiene 33:911–917.

Basch PF. 1986a. Immunocytochemical localization of ecdysteroids in the life history stages of *Schistosoma mansoni*. Comparative Biochemistry and Physiology 83A:199–202.

Basch PF. 1986b. Internal chemical communication within flatworms. Journal of Chemical Ecology 12:1679–1686.

Basch PF. 1988. Schistosoma mansoni: nucleic acid synthesis in immature females from single-sex infections, paired in vitro with intact males and male segments. Comparative Biochemistry and Physiology 90B:389–392.

Basch PF. 1989. Snails as intermediate hosts of fluke diseases. Pp. 762–767 In: R Goldsmith and D Heyneman, Eds. Tropical Medicine and Parasitology. E. Norwalk, CT. Appleton and Lange.

Basch PF. 1990. Why do schistosomes have separate sexes? Parasitology Today 6:160–163.

Basch PF, Basch MN. 1984. Intergeneric reproductive stimulation and parthenogenesis in *Schistosoma mansoni*. Parasitology 89:364–376.

Basch PF, Basch N. 1982. *Schistosoma mansoni*: Scanning electron microscopy of schistosomula, adults and eggs grown in vitro. Parasitology 85:333–338.

Basch PF, Clemens LE. 1989. Schistosoma mansoni: reversible destruction of testes by procarbazine. Comparative Biochemistry and Physiology 93C:397–401.

Basch PF, DiConza JJ. 1974. The miracidium-sporocyst transition in Schistosoma mansoni: Surface changes in vitro with ultrastructural correlation. Journal of Parasitology 60:935–941.

Basch PF, Gupta BC. 1988a. Homosexual pairing in Schistosoma mansoni. International Journal for Parasitology 18:1115–1117.

Basch PF, Gupta BC. 1988b. Immunocytochemical localization of regulatory peptides in six species of trematodes. Comparative Biochemistry and Physiology 90:389–392.

Basch PF, Humbert R. 1981. Cultivation of *Schistosoma mansoni* in vitro. 3. Implantation of cultured worms into mouse mesenteric veins. Journal of Parasitology 67:191–195.

Basch PF, Lie KJ. 1965. Multiple infections of larval trematodes in one snail. Medical Journal of Malaya 20:59.

Basch PF, Lie KJ. 1966a. Infection of single snails with two different trematodes. I. Simultaneous infection and early development of a schistosome and an echinostome. Zeitschrift fur Parasitenkunde 27:252–259.

Basch PF, Lie KJ. 1966b. Infections of single snails with two different trematodes. II. Dual exposures to a schistosome and an echinostome at staggered intervals. Zeitschrift fur Parasitenkunde 27:260–270.

Basch PF, Lie KJ, Heyneman D. 1969. Antagonistic interaction between strigeid and schistosoma sporocysts within a snail host. Journal of Parasitology 55:753–758.

Basch PF, Nicholas C. 1989. Schistosoma mansoni: pairing of male worms with artificial surrogate females. Experimental Parasitology 68:202–207.

Basch PF, O'Toole ML. 1982. Cultivation in vitro of *Schistosomatium douthitti* (Trematoda: Schistosomatidae). International Journal for Parasitology 12:541–545.

Basch PF, Rhine WD. 1983. *Schistosoma mansoni*: reproductive potential of male and female worms cultured in vitro. Journal of Parasitology 69:567–569.

Basch PF, Samuelson J. 1990. Cell biology of schistosomes. I. Ultrastructure and transformations. Pp. 91–106 In: DJ Wyler, Ed. Modern Parasite Biology. New York. WH Freeman and Co.

Bayne CJ. 1980. Humoral factors in molluscan immunity. In: JB Solomon, Ed. Aspects of Developmental and Comparative Immunology—I. New York. Pergamon Press. 113–124.

Bayne CJ. 1982. Recognition and killing of metazoan parasites, particularly in molluscan hosts. Pp. 109–114 In: EL Cooper and MAB Brazier, Eds. Developmental Immunology: Clinical Problems and Aging. UCLA Forum in Medical Sciences 25. New York. Academic Press.

Bayne CJ. 1983. Molluscan Immunobiology. Pp. 407–486 In: ASM Saleuddin, KM Wilbur, Eds. The Mollusca Vol 5, Part 2. New York. Academic Press.

Bayne CJ, Boswell CA, Yui MAA. 1987. Widespread antigenic cross-reactivity between plasma proteins of a gastropod and its trematode parasite. Developmental and Comparative Immunology 11:321–329.

Bayne CJ, Loker ES, Yui MA, Stephens JA. 1984. Immune recognition of Schistosoma mansoni primary sporocysts may require specific receptors on Biomphalaria glabrata hemocytes. Parasite Immunology 6:519–528.

Bayne CJ, Yoshino TP. 1989. Determinants of compatibility in mollusc-trematode parasitism. American Zoologist 29:399–407.

Becker W. 1973. Aktivierung und inaktivierung der Miracidien von Schistosoma mansoni innerhalb der Eischale. Zeitschrift für Parasitenkunde 42:235–242.

Becker W. 1977. Zur Stoffwechselphysiologie der Miracidien von Schistosoma mansoni wahrend ihrer Aktivierung innerhalb der Eischale. Zeitschrift für Parasitenkunde 52:69–79.

Bennett JL, Gianutsos G. 1978. Disulfuram: A compound that selectively induces abnormal egg production and lowers norepinephrine levels in S. mansoni. Biochemical Pharmacology 27:817–820.

Bennett JL, Seed JL, Boff M. 1978. Fluorescent histochemical localization of phenol oxidase in female S. mansoni. Journal of Parasitology 64:941–944.

Berberian DA, Pacquin AOJ, Fantazzi A. 1953. Longevity of Schistosoma haematobium and Schistosoma mansoni: Observations based on a case. Journal of Parasitology 39:517–519.

Berg E. 1953. Effect of castration in male mice on Schistosoma mansoni. Proceedings of the Society for Experimental Biology and Medicine 83:83–85.

Berg E. 1957. Effects of castration and testosterone in male mice on Schistosoma mansoni. Transactions of the Royal Society of Tropical Medicine and Hygiene 51:353–358.

Bidinger PD, Crompton DWT. 1989. A possible focus of schistosomiasis in Andhra Pradesh, India. Transactions of the Royal Society of Tropical Medicine and Hygiene 83:526.

Bjorneboe A, Frandsen F. 1979. A comparison of characteristics of two strains of Schistosoma intercalatum Fisher, 1934 in mice. Journal of Helminthology 53:195–203.

Blair DM. 1966a. The occurrence of terminal spined eggs, other than those of Schistosoma haematobium, in human beings in Rhodesia. Central African Journal of Medicine 12:103–109.

Blair DM. 1966b. Heads or tails: Where does the head of the miracidium lie in the schistosome egg? Central African Journal of Medicine 12:146–148.

Blanton RE, Davis AH, Rottman F, Maurer R, Mahmoud AA. 1985. Three developmentally regulated genes identified by clones from an adult Schistosoma mansoni cDNA expression library. Transactions of the Association of American Physicians 98:101–106.

Bloch EH. 1980. In vivo microscopy of schistosomiasis II. Migration of *Schistosoma mansoni* in the lungs, liver, and intestine. American Journal of Tropical Medicine and Hygiene 29:62–70.

Bobek LA, LoVerde PJ, Rekosh DM. 1989. Schistosoma haematobium: analysis of eggshell protein genes and their expression. Experimental Parasitology 68: 17–30.

Bobek L, Rekosh DM, Van Keulen H, LoVerde PT. 1986. Characterization of a female-specific cDNA derived from a developmentally regulated messenger RNA in the human blood fluke *Schistosoma mansoni*. Proceedings of the National Academy of Sciences, U.S.A. 83:5544–5548.

Boctor FN, Nash TE, Cheever AW. 1979. Isolation of a polysaccharide antigen from *Schistosoma mansoni* eggs. Journal of Immunology 122:39–43.

Bone LW. 1982. Reproductive chemical communication of helminths. I. Platyhelminthes. International Journal of Invertebrate Reproduction 5:261–268.

Bourns TKR, Ellis JC, Rau ME. 1973. Migration and development of *Trichobilharzia ocellata* (Trematoda: Schistosomatidae) in its duck host. Canadian Journal of Zoology 51:1021–1030.

Braga VMM, Tavares CAP, Rumjanek FD. 1989. Protein characterization of sexually mature and immature forms of Schistosoma mansoni. Comparative Biochemistry and Physiology 94B:427–433.

Brémond P, Théron A, Rollinson D. 1989. Hybrids between Schistosoma mansoni and Schistosoma rodhaini: Characterization by isoelectric focusing of six enzymes. Parasitology Research 76:138–145.

Briggs MH. 1972. Metabolism of steroid hormones by schistosomes. Biochimica et Biophysica Acta 280:481–485.

Brindley PJ, Lewis FA, McCutchan TF, Bueding E, Sher A. 1989. A genomic change associated with the development of resistance to hycanthone in Schistosoma mansoni. Molecular and Biochemical Parasitology 36:243–252.

Brink LD, McLaren DJ, Smithers SR. 1977. Schistosoma mansoni: A comparative study of artificially transformed schistosomula and schistosomula recovered after cercarial penetration of mouse skin. Parasitology 74:73–86.

Brooks CP. 1953. A comparative study of Schistosoma mansoni in Tropicorbis havanensis and Australorbis glabratus. Journal of Parasitology 39:159–165.

Brown JN, Smith TM, Doughty BL. 1977. Phospholipid, fatty acid and lipid class composition of eggs from the human blood fluke *Schistosoma japonicum*. Federation Proceedings 36:1159.

Browne H, Schulert AR. 1963. Deposition of Sb124-labeled Astiban (R) in animals with bilharziasis. Federation Proceedings 22:666.

Bruce JI, Dias LC, Liang YS, Coles GC. 1987. Drug resistance in schistosomiasis: a review. Memorias do Instituto Oswaldo Cruz 82 Suppl 4:143–150.

Bruce JI, Ruff MD, Belusko RJ, Werner JK. 1972. *Schistosoma mansoni* and *Schistosoma japonicum*: Utilization of amino acids. International Journal for Parasitology 2:425–430.

Bruckner DA, Schiller EL. 1974. Some biological characteristics of Liberian and Puerto Rican strains of *Schistosoma mansoni*. Journal of Parasitology 60: 551–552.

Bruijning CFA. 1968. An abnormal *Schistosoma mansoni* egg. Acta Leidensia 36:23–25.

Brumpt E. 1930a. Cycle évolutif complet de *Schistosoma bovis*. Annales de Parasitologie Humaine et Comparée 8:17–50.

Brumpt E. 1930b. Localisation vesicale experimentale d'oeufs de *Schistosoma mansoni* chez une souris. Localisations vesicales à *S. mansoni* et rectales à *S. haematobium* chez l'homme. Annales de Parasitologie Humaine et Comparée 8:298–308.

Brumpt E. 1930c. La ponte des schistosomes. Annales de Parasitologie Humaine et Comparée 263–297.

Brumpt E. 1931. Description de deux bilharzies de mammifères africains, *Schistosoma curassoni* sp. inquir. et *Schistosoma rodhaini*, n. sp. Annales de Parasitologie Humaine et Comparée 9:325–338.

Brumpt E. 1936. Action des hôtes définitifs sur l'évolution et sur la selection des sexes de certaines helminthes héberges par eux. Experiences sur des schistosomes. Annales de Parasitologie Humaine et Comparée 14:543–551.

Brumpt E. 1949. Précis de Parasitologie. Paris. Masson et Cie.

Bryant C, Flockhart HA. 1986. Biochemical strain variation in parasitic helminths. Advances in Parasitology 25:275–319.

Bueding E, Koletsky S. 1950. Content and distribution of glycogen in *Schistosoma mansoni*. Proceedings of the Society for Experimental biology and Medicine 73:594–596.

Bueding E, Schiller EL, Bourgeois J. 1967. Some physiological, biochemical, and morphologic effects of tris(p–aminophenyl) carbonium salts (TAC) on *Schistosoma mansoni*. American Journal of Tropical Medicine and Hygiene 16:500–515.

Burchard GD, Kern P. 1985. Probable hybridization between *S. intercalatum* and *S. haematobium* in western Gabon. Tropical and Geographic Medicine 37:119–123.

Burden CS, Ubelaker JE. 1981. *Schistosoma mansoni* and *Schistosoma haematobium*: differences in development. Experimental Parasitology 51:28–34.

Buttner A. 1950. Curieux cas d'hermaphroditisme chez une souche Africaine de *Schistosoma mansoni*. Comptes Rendus des Séances de l'Académie des Sciences 230:1420–1422.

Buttner A. 1951. Labilité particulière du sexe chez *Schistosoma mansoni* (Platyhelminthe, Trematode). Essai d'interprétation. Annales de Parasitologie Humaine et Comparée 25:297–307.

Byram JE, Senft AW. 1979. Structure of the schistosome eggshell: amino acid analysis and incorporation of labelled amino acids. American Journal of Tropical Medicine and Hygiene 28:539–547.

Campbell WC, Cuckler AC. 1967. Inhibition of egg production of *Schistosoma mansoni* in mice treated with nicarbazin. Journal of Parasitology 53:977–980.

Cao HM, Wang YF, Long S. 1982. A study of ultrastructure of egg shell of *Schistosoma japonicum* I. Transmission electron microscopic observation on *S. japonicum* egg. Annales de Parasitologie Humaine et Comparée 57:345–352.

Capron A, Vernes A. 1968. Etude des précipitines seriques au cours des infections unisexuées experimentales par Schistosoma mansoni. Structure antigenique de l'oeuf de S. mansoni. Revue d'immunologie et de Thérapie Antimicrobienne 32:209–222.

Carmichael AC. 1984. Phylogeny and Historical Biogeography of the Schistosomatidae. PhD Dissertation, Michigan State University, East Lansing, Michigan, 257 p.

Carmichael AC, Muchlinski AE. 1980. Survival of *Schistosomatium douthitti* during hibernation in the natural host, Zapus hudsonicus. Journal of Parasitology 66:365–366.

Carney WP, Brown RJ, Van Peenen PFD, Purnomo, Ibrahim B, Koesharjono CR. 1977. Schistosoma incognitum from Cikurai, West Java, Indonesia. International Journal for Parasitology 7:361–365.

Cesari IM. 1974. *Schistosoma mansoni*: distribution and characteristics of acid and alkaline phosphatases. Experimental Parasitology 36:405–414.

Cesari IM, Rodriguez M, McLaren DJ. 1981. Proteolytic enzymes in adult male and female worms of *Schistosoma mansoni*. Actas Cientificas Venezolanas 32:324–329.

Chabasse D, Bertrand G, Leroux JP, Gauthey N, Hocquet P. 1985. Bilharziose à *Schistosoma mansoni* évolutiv decouverte 37 ans après l'infestation. Bulletin de la Société de Pathologie Exotique 78:643–647.

Chappell LH. 1974. Methionine uptake of larval and adult *Schistosoma mansoni*. International Journal for Parasitology 4:361–369.

Cheever AW. 1968. A quantitative post-mortem study of *Schistosomiasis mansoni* in man. American Journal of Tropical Medicine and Hygiene 17:38–64.

Cheever AW. 1986. Density-dependent fecundity in Schistosoma mansoni infections in man: a reply (letter). Transactions of the Royal Society of Tropical Medicine and Hygiene 86:991–992.

Cheever AW. 1988. Lack of effect of unmated schistosomes on the fecundity of mated worm pairs. Journal of Parasitology 74:249–252.

Cheever AW, Dunn MA, Dean DA, Duvall RH. 1983. Differences in hepatic fibrosis in ICR, C3H and C57BL/6 mice infected with Schistosoma mansoni. American Journal of Tropical Medicine and Hygiene 32:1364–1369.

Cheever AW, Duvall RH. 1974. Single and repeated infections of grivet monkeys with Schistosoma mansoni: Parasitological and pathological observations over a 31-month period. American Journal of Tropical Medicine and Hygiene 23:884–894.

Cheever AW, Duvall RH. 1982. *Schistosoma japonicum*: migration of adult worm pairs within the mesenteric veins of mice. Transactions of the Royal Society of Tropical Medicine and Hygiene 76:641–645.

Cheever AW, Duvall RH. 1987. Variable maturation and oviposition by female Schistosoma japonicum in mice: the effects of irradiation of the host prior to infection. American Journal of Tropical Medicine and Hygiene 37:562–569.

Cheever AW, Duvall RH, Hallack TA Jr, Minker RG, Malley JD, Malley KG. 1987. Variation of hepatic fibrosis and granuloma size among mouse strains infected with Schistosoma mansoni. American Journal of Tropical Medicine and Hygiene 37:85–97.

Cheever AW, Duvall RH, Kuntz RE, Huang TC, Moore JA. 1988. Resistance of capuchin monkeys to re-infection with Schistosoma mansoni. Transactions of the Royal Society of Tropical Medicine and Hygiene 82:112–114.

Cheever AW, Kamel IA, Elwi AM, Mosimann JG, Danner R. 1977. *Schistosoma mansoni* and *Schistosoma haematobium* infections in Egypt II. Quantitative parasitological findings at necropsy. American Journal of Tropical Medicine and Hygiene 26:702–716.

Cheever AW, Powers KG. 1969. *Schistosoma mansoni* infection in rhesus monkeys:

changes in egg production and egg distribution in prolonged infections in intact and splenectomized monkeys. Annals of Tropical Medicine and Parasitology 63:83–93.

Cheever AW, Powers KG. 1972. Schistosoma mansoni infection in rhesus monkeys: Comparison of the course of heavy and light infections. Bulletin of the World Health Organization 46:301–309.

Cheever AW, Torkey AH, Shirbiney M. 1975. The relation of worm burden to passage of *Schistosoma haematobium* eggs in the urine of infected patients. American Journal of Tropical Medicine and Hygiene 24:284–288.

Cheng TC. 1979. Cellular immunity in molluscs: with emphasis on the intermediate hosts of human-infecting schistosomes. Institute of Animal Resources News 22:9–16.

Chernin E. 1966. Transplantation of larval Schistosoma mansoni from infected to uninfected snails. Journal of Parasitology 52:473–482.

Chien K, Liu S-t. 1960. A note on hermaphroditism of female *Schistosoma japonicum*. Tung Wu Hsueh Pao 12:9–11.

Chiu J-K, Kao C-T. 1973. Studies of host-parasite relationships of Ilan strain of *Schistosoma japonicum* in animals. Chinese Journal of Microbiology 6:72–79.

Chu GWTC, Cutress CE. 1954. Austrobilharzia variglandis (Miller and Northrup, 1926) Penner 1953 (Trematoda: Schistosomatidae) in Hawaii with notes on its biology. Journal of Parasitology 40:515–524.

Cicchini T, Passai S, Simoni L, Messeri E, Belli M. 1977. L'importanzia delle frazioni lipidiche cutanee di uomo, di topo e di ratto nella penetrazione "in vitro" delle cercarie di "Schistosoma mansoni." Annali Sclavo 19:1033–1042.

Cioli D, Knopf PM. 1980. A study of the mode of action of hycanthone against *Schistosoma mansoni* in vivo and in vitro. American Journal of Tropical Medicine and Hygiene 29:220–226.

Cioli D, Knopf PM, Senft AW. 1977. A study of *Schistosoma mansoni* transferred into permissive and nonpermissive hosts. International Journal for Parasitology 7:293–297.

Cioli D, Pica Mattoccia L. 1984. Genetic analysis of hycanthone resistance in *Schistosoma mansoni*. American Journal of Tropical Medicine and Hygiene 33:80–88.

Clark WC. 1974. Interpretation of life history pattern in the digenea. International Journal for Parasitology 4:115–123.

Clegg JA. 1959. Development of sperm by *Schistosoma mansoni* cultured in vitro. Bulletin of the Research Council of Israel 8E:1–6.

Clegg JA. 1965b. In vitro cultivation of *Schistosoma mansoni*. Experimental Parasitology 16:133–147.

Clegg JA, Morgan, J. 1966. The lipid composition of the lipoprotein membranes on the egg-shell of *Fasciola hepatica*. Comparative Biochemistry and Physiology 18:573–588.

Clemens LE, Basch PF. 1989a. Effects of transferrin and mammalian growth factors on development of Schistosoma mansoni in vitro. Journal of Parasitology 75:417–421.

Clemens LG, Basch PF. 1989b. Schistosoma mansoni: insulin independence. Experimental Parasitology 68:223–229.

Clough ER. 1981. Morphology and reproductive organs and oogenesis in bisexual and unisexual transplants of mature *Schistosoma mansoni* females. Journal of Parasitology 67:535–539.

Cohen LM, Eveland LK. 1984. *Schistosoma mansoni*: Long-term maintenance of clones by the microsurgical transplantation of sporocysts. Experimental Parasitology 57:15–19.

Cohen LM, Eveland LK. 1988. Schistosoma mansoni: characterization of clones maintained by the microtransplantation of sporocysts. Journal of Parasitology 79:963–969.

Coles GC. 1973. The metabolism of schistosomes: a review. International Journal of Biochemistry 4:319–337.

Coles GC. 1984. Recent advances in schistosome biochemistry. Parasitology 89: 603–637.

Coles GC, Thurston JP. 1970. Testes number in East African Schistosoma mansoni. Journal of Helminthology 44:69–73.

Colley DG, Wikel SK. 1974. Schistosoma mansoni: Simplified method for the production of schistosomules. Experimental Parasitology 35:44–51.

Combes C. 1985. L'analyse de la compatibilité Schistosomes/Mollusques Vecteurs. Bulletin de la Société de Pathologie Exotique et ses Filiales 78:742–746.

Combescot C, Barrabes A, Demaret J. 1971. Rôle de l'oestradiol dans la parasitose experimentale à Schistosome mansoni chez le hamster doré femelle Cricetus auratus. Comptes Rendus des Seances de la Societe de Biologie et de ses Filiales (Paris) 164:1623–1625.

Combescot C, Barrabes A, Reynouard F. 1974. Effet protecteur de l'oestradiol au cours de la parasitose expérimentale à Schistosoma mansoni chez le hamster doré femelle castré; importance de la barriere cutanée. Annales de Parasitologie Humaine et Comparée 49:185–189.

Cook GC, Bryceson ADM. 1988. Longstanding infection with Schistosoma mansoni. Lancet 1:127.

Cooper LA, Lewis FA, File-Emperador S. 1989. Re-establishing a life cycle of Schistosoma mansoni from cryopreserved larvae. Journal of Parasitology 75:353–356.

Cornford EM. 1985. *Schistosoma mansoni, S. japonicum,* and *S. haematobium:* Permeability to acidic amino acids and effect of separated and unseparated adults. Experimental Parasitology 59:355–363.

Cornford EM, Diep CP, Rowley GA. 1983. *Schistosoma mansoni, S. japonicum, S. haematobium:* Glycogen content and glucose uptake in parasites from fasted and control hosts. Experimental Parasitology 56:397–408.

Cornford EM, Fitzpatrick AM. 1985. The mechanism and rate of glucose transfer from male to female schistosomes. Molecular and Biochemical Parasitology 17:131–141.

Cornford EM, Fitzpatrick AM. 1987. Comparative glucose utilization rates in separated and mated schistosomes. Experimental Parasitology 64:448–457.

Cornford EM, Fitzpatrick AM, Quirk TL, Diep CP, Landaw EM. 1988. Tegumental glucose permeability in male and female Schistosoma mansoni. Journal of Parasitology 74:116–128.

Cornford EM, Huot M. 1981. Glucose transfer from male to female schistosomes. Science 213:1269–1271.

Cornford EM, Oldendorf WH. 1979. Transintegumental uptake of metabolic substrates in male and female *Schistosoma mansoni.* Journal of Parasitology 65:357–363.

Correa-Oliveira R, James SL, McCall D, Sher A. 1986. Identification of a genetic locus Rsm-1, controlling protective immunity against Schistosoma mansoni. Journal of Immunology 137:2014–2019.

Correa-Oliveira R, James SL, Sher A. 1988. Genetic complementation of defects in vaccine-induced immunity against Schistosoma mansoni in P- and A-strain inbred mice. Infection and Immunity 56:708–710.

Cort WW. 1919. Notes on the eggs and miracidia of the human schistosomes. University of California Publications in Zoology 18:509–519.

Cort WW. 1921a. The development of the Japanese blood-fluke, Schistosoma japonicum Katsurada, in its final host. American Journal of Hygiene 1:1–38.

Cort WW. 1921b. Sex in the family Schistosomatidae. Science 53:226–228.

Cousin CE, Stirewalt MA, Dorsey CH. 1981. S. mansoni: ultrastructure of early transformation of skin- and shear-pressure derived schistosomules. Experimental Parasitology 51:341–365.

Crabtree JE, Wilson RA. 1980. Schistosoma Mansoni: a scanning electron microscope study of the developing schistosomulum. Parasitology 81:553–564.

Crabtree JE, Wilson RA. 1986. Schistosoma mansoni: an ultrastructural examination of pulmonary migration. Parasitology 92:343–354.

Cram EB, Files VS, Jones MF. 1947. Experimental molluscan infection with Schistosoma mansoni and Schistosoma haematobium. National Institute of Health Bulletin 189:81–94.

Damian RT, Chapman RW. 1983. The fecundity of Schistosoma mansoni in baboons, with evidence for a sex ratio effect. Journal of Parasitology 69:987–989.

Damian RT, Rawlings CA, Bosshardt SC. 1986. The fecundity of Schistosoma mansoni in chronic nonhuman primate infections before and after transplantation into naive hosts. Journal of Parasitology 72:741–747.

Davies P, Jackson H. 1970. Experimental studies on the chemosterilization of Schistosoma mansoni. Parasitology 61:167–176.

Davis AH, Blanton R, Klich P. 1985. Stage and sex specific differences in actin gene expression in Schistosoma mansoni. Molecular and Biochemical Parasitology 17:289–298.

Dean DA. 1983. A review. Schistosoma and related genera: Acquired resistance in mice. Experimental Parasitology 55:1–104.

Dean DA, Mangold BL, Kassim OO, Von Lichtenberg F. 1987. Sites and mechanisms of schistosome elimination. Memorias do Instituto Oswaldo Cruz 82 Suppl 4:31–37.

de Boissezon B, Jelnes JE. 1982. Isoenzyme studies on cercariae from monoinfections and adult worms of Schistosoma mansoni (10 isolates) and S. rodhaini (one isolate) by horizontal polyacrylamide gel electrophoresis and staining of eight enzymes. Zeitschrift für Parasitenkunde 67:185–196.

de Carneri I. 1963. Coppie omosessuali di maschi di Schistosoma mansoni nel topo. Rivissta di Parassitologia 24:179–183.

de Gentile L, Fayad M, Denis P, Lecestre MJ. 1988. Localisation erratique et longevité exceptionelle de Schistosoma mansoni. Journal de Urologie (Paris) 94:163–165.

Dei-Cas E, Dhainaut-Courtois N, Vernes A. 1980. Contribution a l'étude du systeme nerveux des formes adultes et larvaires de Schistosoma mansoni Sambon, 1907 (Trematoda Digenea). I. Aspects morphologiques: anatomie, histologie et ultrastructure chez la forme adulte. Annales de Parasitologie Humaine et Comparée 55:69–86.

De Meillon B, Paterson S. 1958. Experimental bilharziasis in animals. VII. Effect of a low-protein diet on bilharziasis in white mice. South African Medical Journal 32:1086–1088.

Den Hollander JE, Erasmus DE. 1984. *Schistosoma mansoni*: DNA synthesis in males and females from mixed and single-sex infections. Parasitology 88: 463–476.

Den Hollander JE, Erasmus DE. 1985. *Schistosoma mansoni*: Male stimulation and DNA synthesis by the female. Parasitology 91:449–457.

Deters DL, Nollen PM. 1976. The presence of oocytes in the testes of *Schistosoma haematobium*. Journal of Parasitology 62:324–325.

DeWitt WB. 1954. Susceptibility of snail vectors to geographic strains of Schistosoma japonicum. Journal of Parasitology 40:453–456.

DeWitt WB. 1957a. Experimental schistosomiasis mansoni in mice maintained on nutritionally deficient diets. I. Effects of a Torula yeast ration deficient in factor 3, vitamin E and cystine. Journal of Parasitology 43:119–128.

DeWitt WB. 1957b. Experimental schistosomiasis mansoni in mice maintained on nutritionally deficient diets. II. Survival and development of *Schistosoma mansoni* in mice maintained on a Torula yeast diet deficient in factor 3, vitamin E, and cystine. Journal of Parasitology 43:129–135.

DeWitt WB, Oliver-Gonzalez J, Medina E. 1964. Effects of improving the nutrition of malnourished people infected with *Schistosoma mansoni*. American Journal of Tropical Medicine and Hygiene 13:25–35.

Dias LC, Bruce JI, Coles GC. 1988. Variation in response of Schistosoma mansoni strains to schistosomicides. Revista do Instituto de Medicina Tropical de Sao Paulo 30:81–85.

Dias LC, Olivier CE. 1985. Stability of Schistosoma mansoni progeny to antischistosomal drugs. Revista do Instituto de Medicina Tropical de Sao Paulo 27: 186–189.

DiConza JJ, Basch PF. 1974. Axenic cultivation of *Schistosoma mansoni* daughter sporocysts. Journal of Parasitology 60:757–763.

DiConza JJ, Hansen EL. 1972. Multiplication of transplanted Schistosoma mansoni daughter sporocysts. Journal of Parasitology 58:181–182.

Dinnik JA, Dinnik NN. 1965. The schistosomes of domestic ruminants in eastern Africa. Bulletin of Epizootic Diseases of Africa 13:341–359.

Dissous C, Dissous CA, Capron A. 1986. Stimulation of Schistosoma mansoni miracidia by a 80 KDa glycoprotein from Biomphalaria glabrata. Molecular and Biochemical Parasitology 21:203–209.

Doughty BA. 1975. Vitellogenesis and egg shell formation in *Schistosoma mansoni*. Abstracts, U. S.-Japan Cooperative Medical Sciences Program Joint Conference on Parasitic Diseases, Bethesda, Maryland, October 27–29. Pp. 51–53.

Dutt SC. 1967. Studies on *Schistosoma nasale* Rao, 1933 I. Morphology of the adults, egg, and larval stages. Indian Journal of Veterinary Science and Animal Husbandry 37:249–262.

Dutt SC, Srivastava HD. 1968. Studies on *Schistosoma nasale* Rao, 1933 II. Mulluscan and mammalian hosts of the blood-fluke. Indian Journal of Veterinary Science and Animal Husbandry 38:210–216.

Ebrahimzadeh A. 1966. Histologisch Untersuchungen uber den Feinbau des oogentop bei Digenen Trematoden. Zeitschrift für Parasitenkunde 27:127–168.

El-Gindy MS. 1951. Post-cercarial development of *Schistosomatium douthitti* (Cort, 1914) Price 1931 in mice, with special reference to the genital system (Schistosomatidae-Trematoda). Journal of Morphology 89:151–185.

El-Hawy AM, El-Tayees S, Solem SN, El-Nasr MS, Sabry H. 1989. HLA typing intestinohepatic schistosomiasis. Journal of the Egyptian Society of Parasitology 19:417–425.

Elku-Natey DT, Wüest J, Swiderski Z, Striebel HP, Huggel H. 1985. Comparative scanning electron microscope (SEM) study of miracidia of four human schistosome species. International Journal for Parasitology 15:33–42.

El Raggal M. 1959. Wird das Geschlecht der Schistosomen durch das der Wirtschnecke beeinflusst?. Zeitschrift für Tropenmedizin und Parasitologie 10: 66–70.

Elsaghier AAF, Knopf PM, Mitchell GF, McLaren DJ. 1989. Schistosoma mansoni: evidence that "non-permissiveness" in 129/Ola mice involves worm relocation and attrition in the lungs. Parasitology 99:365–375.

Elsaghier AAF, McLaren DJ. 1989. Schistosoma mansoni: evidence that vascular abnormalities correlate with the "non-permissiveness" trait in 129/Ola mice. Parasitology 99:377–381.

el-Sherbeini M, Bostian KA, Knopf PM. 1990. Schistosoma mansoni: cloning of antigen gene sequences in Escherichia coli. Experimental Parasitology 70: 72–84.

Erasmus DA. 1972. The Biology of Trematodes. New York. Crane, Russak. 312 p.

Erasmus DA. 1973. A comparative study of the reproductive system of mature, immature and "unisexual" female Schistosoma mansoni. Parasitology 67: 165–183.

Erasmus DA. 1975a. The subcellular localization of labelled tyrosine in the vitelline cells of Schistosoma mansoni. Zeitschrift für Parasitenkunde 46:75–81.

Erasmus DA. 1975b. Schistosoma mansoni: development of the vitelline cell, its role in drug sequestration, and changes induced by Astiban. Experimental Parasitology 38:240–256.

Erasmus DA, Davies TW. 1979. Schistosoma mansoni and S. haematobium: calcium metabolism of the vitelline cell. Experimental Parasitology 47:91–106.

Erasmus DA, Popiel I. 1980. Schistosoma mansoni: drug induced changes in the cell population of the vitelline gland. Experimental Parasitology 50:171–187.

Erasmus DA, Popiel I, Shaw JR. 1982. A comparative study of the vitelline cell in Schistosoma mansoni, S. haematobium, S. japonicum, and S. mattheei. Parasitology 84:283–287.

Erasmus DA, Shaw JR. 1977. Egg production in Schistosoma mansoni. Transactions of the Royal Society of Tropical Medicine and Hygiene 71:289.

Ercoli N, Payares G, Nunez D. 1985. Schistosoma mansoni: neurotransmitters and the mobility of cercariae and schistosomules. Experimental Parasitology 59:204–216.

Evans AS, Stirewalt 1957. Variations in infectivity of cercariae of Schistosoma mansoni. Experimental Parasitology 1:19–33.

Eveland LK, Fried B, Cohen LM. 1982. Schistosoma mansoni: Adult worm chemoattraction, with and without barriers. Experimental Parasitology 54:271–276.

Eveland LK, Fried B, Cohen LM. 1983. Schistosoma mansoni: Adult worm chemoattraction with barriers of specific molecular weight exclusions. Experimental Parasitology 56:255–258.

Eveland LK, Haseeb M. 1989. Schistosoma mansoni: Onset of chemoattraction in developing worms. Experientia 45:309–310.

Evers P, Jackson TF, Dettman C, Sapsford C. 1983. A comparative scanning electron microscope study of the teguments of adult male Schistosoma mansoni, S.

margrebowiei, and *S. leiperi.* Scanning Electron Microscopy 1983 (Part 1):215–220.

Fairley NH. 1920. A comparative study of experimental bilharziasis in monkeys contrasted with the hitherto described lesions in man. Archives of Pathology and Bacteriology 23:289–314.

Fairley NH, Mackie FP, Jasudasan F. 1930. Studies on *S. spindale.* V. The guinea pig as a host for male schistosomes. Indian Medical Research Memoirs 17:61–72.

Fanning MM, Kazura JM. 1984. Genetic-linked variation in susceptibility of mice to *schistosomiasis mansoni.* Parasite Immunology 6:95–103.

Farley J. 1971. A review of the family Schistosomatidae excluding the genus Schistosoma from mammals. Journal of Helminthology 45:289–320.

Faust EC. 1927. The possible presence of male schistosomes alone in experimental natural infections. Journal of Parasitology 14:62–63.

Faust EC, Jones CA, Hoffman WA. 1934. Studies on schistosomiasis in Puerto Rico. III Biological studies. 2. The mammalian phase of the life cycle. Puerto Rican Journal of Public Health and Tropical Medicine 10:133–196.

Faust EC, Meleney HE. 1924. Studies on *Schistosomiasis japonica.* American Journal of Hygiene Monograph Series 3: 339 Pp.

Feiler W, Haas W. 1988a. Tricholbilharzia ocellata: chemical stimuli of duck skin for cercarial attachment. Parasitology 96:507–517.

Feiler W, Haas W. 1988b. Host-finding in Tricholbilharzia ocellata cercariae: swimming and attachment to the host. Parasitology 96:493–505.

Files VS. 1951. A study of the vector-parasite relationships in Schistosoma mansoni. Parasitology 41:264–269.

Files VS, Cram EB. 1949. A study on the comparative susceptibility of snail vectors to strains of Schistosoma mansoni. Journal of Parasitology 35:555–560.

Fisher AC. 1934. A study of the schistosomiasis of the Stanleyville District of the Belgian Congo. Transactions of the Royal Society of Tropical Medicine and Hygiene 28:277–306.

Fletcher M, LoVerde PT, Kuntz RE. 1981a. Electrophoretic differences between *Schistosoma mansoni* and *S. rodhaini.* Journal of Parasitology 67:593–595.

Fletcher M, LoVerde PT, Richards CS. 1981b. *Schistosoma mansoni:* Electrophoretic characterization of strains selected for different levels of infectivity to snails. Experimental Parasitology E2:362–370.

Fletcher M, LoVerde PT, Woodruff D. 1981c. Genetic variation in *Schistosoma mansoni:* enzyme polymorphisms in populations from Africa, Southwest Asia, South America, and the West Indies. American Journal of Tropical Medicine and Hygiene 30:406–421.

Foley M, Kusel JR, Garland PB. 1988. Changes in the organization of the surface membranes upon transformation of cercariae to schistosomula of the helminth parasite Schistosoma mansoni. Parasitology 96:85–97.

Ford JW, Blankespoor HD. 1979. Scanning electron microscopy of the eggs of three human schistosomes. International Journal for Parasitology 9:141–145.

Foster R, Cheetham BL. 1973. Studies with the schistosomicide Oxamniquine (UK-4271) 1. Activity in rodents and in vitro. Transactions of the Royal Society of Tropical Medicine and Hygiene 67:674–684.

Foster R, Cheetham BL, Mesmer ET. 1968. The distribution of *Schistosoma mansoni* (Sambon, 1907) in the vertebrate host, and its determination by perfusion. Journal of Tropical Medicine and Hygiene 71:139–145.

Fournier A, Théron A. 1985. Sectorisation morpho-anatomique et fonctionelle du sporocyste-fils de Schistosoma mansoni. Zeitschrift für Parasitenkunde 71:325–336.

Frandsen F. 1977. Investigation of the unimiracidial infection of Schistosoma intercalatum in snails and the infection of the final host using cercariae of one sex. Journal of Helminthology 51:5–10.

Frandsen F. 1979. Discussion of the relationships between Schistosoma and their intermediate hosts, assessment of the degree of host-parasite compatibility and evaluation of schistosome taxonomy. Zeitschrift für Parasitenkunde 58:275–296.

Fried B. 1962. Growth of Philophtalmus sp. in the eyes of chicks. Journal of Parasitology 48:395–399.

Fried B. 1986. Chemical communication in hermaphroditic digenetic trematodes. Journal of Chemical Ecology 12:1659–1677.

Fried B, Shenko FJ, Eveland LK. 1983. Densitometric thin-layer chromatographic analyses of cholesterol in Schistosoma mansoni (Trematoda) adults and their excretory-secretory products. Journal of Chemical Ecology 9:1483–1489.

Fripp PJ. 1967. On the morphology of Schistosoma rodhaini (Trematoda, Digenea, Schistosomatidae). Journal of Zoology (London) 151:433–452.

Fripp PJ. 1968. Some observations on the behaviour of the Kampala strain of Schistosoma rodhaini Brumpt in the laboratory. South African Journal of Medical Science 33:21–20.

Fritsch G. 1988. Zur Anatomie der Bilharzia haematobium (Cobbold). Archiv für Mikroskopische Anatomie 31:192–223.

Fujinami A. 1907. Weitere Mitteilungen über die "Katayama-Krankheit" (die durch Schistosomum japonicum hervorgerufene endemische Krankheit in Katayama). Kyoto Igaku Zassi 4:15–22.

Fujinami A, Nakamura H. 1909. Neue Untersuchungen über die japanischer Schistosomum-Krankheit (Katayama-Krankheit). Invasions pforte und Wachstum des Parasiten. Die zu dieser Krankheit disponierten Tiere. Kyoto Igaku Zasshi 6:228–252.

Fulford AJ, Yeang F. 1988. Analysis of worm burdens in experimental schistosomiasis. Parasitology 97:303–315.

Gadgil RK. 1963. Human schistosomiasis in India. Indian Journal of Medical Research 51:244–251.

Gadgil RK, Shah SN. 1956. Human schistosomiasis in India. Establishing the life-cycle in the laboratory. Indian Journal of Medical Research 44:577–580.

Garcia EG. 1976. The Biology of Schistosoma japonicum, Philippine strain: a review. Southeast Asian Journal of Tropical Medicine and Public Health 7:190–196.

Garcia EG, Tiu W, Mitchell GF. 1983. Innate resistance to Schistosoma japonicum in a proportion of 129/J mice. Journal of Parasitology 69:613–615.

Gazzinelli G, Montesanto MA, Correa-Oliveira R, Lima MS, Katz N, Rocha RS, Colley DG. 1987. Immune response in different clinical groups of Schistosomiasis patients. Memorias do Instituto Oswaldo Cruz 82 Suppl 4:95–100.

Georgi JR, Wade SE, Dean DA. 1986. Attrition and temporal distribution of Schistosoma mansoni schistosomula in laboratory mice. Parasitology 93:55–70.

Giovannola A. 1936. Unisexual infection with Schistosoma mansoni. Journal of Parasitology 22:289–290.

Göbel E, Pan JP. 1985. Ultrastructure of the daughter sporocyst and developing cercaria of Schistosoma japonicum in experimentally infected snails, Oncomelania hupensis hupensis. Zeitschrift fur Parasitenkunde 71:227–240.

Goddard MJ, Jordan P. 1980. On the longevity of *Schistosoma mansoni* in man on St. Lucia, West Indies. Transactions of the Royal Society of Tropical Medicine and Hygiene 79:185–191.

Goff WL, Ronald NC. 1981. Certain aspects of the biology and life cycle of *Heterobilharzia americana* in east central Texas. American Journal of Veterinary Research 42:1775–1777.

Gönnert R. 1947. Zur Frage der Wirkungsmechanismus von Miracil. Naturwissenschaften 34:347–348.

Gönnert R. 1949a. Die Struktur der Körperoberflache von *Bilharzia mansoni*. Zeitschrift für Tropenmedizin und Parasitologie 1:105–112.

Gönnert R. 1949b. Über rudimentare weibliche Geschlechtsanlagen bei *Bilharzia mansoni* Männchen. Zeitschrift für Tropenmedizin und Parasitologie 1:272–279.

Gönnert R. 1955a. Schistosomiasis Studien. I. Beitrage zur Anatomie und Histologie von *Schistosoma mansoni*. Zeitschrift für Tropenmedizin und Parasitologie 6:18–33.

Gönnert R. 1955b. Schistosomiasis Studien. II Über die Eibilding bei *Schistosoma mansoni* und das Schicksal der Eier in Wirtsorganismus. Zeitschrift für Tropenmedizin und Parasitologie 6:33–52.

Gönnert R. 1962. Histologische untersuchungen uber den Feinbau der Eibildungsstatte (Oogentop) von *Fasciola hepatica*. Zeitschrift für Parasitenkunde 21:475–492.

Gordon RM, Davey TH, Peaston H. 1934. The transmission of human bilharziasis in Sierra Leone, with an account of the life-cycle of the schistosomes concerned, *S. mansoni* and *S. haematobium*. Annals of Tropical Medicine and Parasitology 28:323–418.

Grant CT, Senft AW. 1971. Schistosome proteolytic enzyme. Comparative Biochemistry and Physiology 38:663–678.

Granzen M, Haas W. 1986. The chemical stimuli of human skin surface for the attachment response of Schistosoma mansoni cercariae. International Journal for Parasitology 16:575–579.

Greer GJ, Kitikoon V, Lohachit C. 1989. Morphology and life cycle of Schistosoma sinensium Pao, 1959, from Northwest Thailand. Journal of Parasitology 75:98–101.

Greer GJ, Ow-Yang CK, Yong HS. 1988. Schistosoma malayensis n. sp.: a Schistosoma japonicum-complex schistosome from Peninsular Malaysia. Journal of Parasitology 74:471–480.

Gresson RAR. 1964. Oogenesis in the hermaphroditic digenea (Trematoda). Parasitology 54:409–421.

Grossman AI, McKenzie R, Cain GD. 1980. Sex heterochromatin in *Schistosoma mansoni*. Journal of Parasitology 66:368–370.

Grossman AI, Short RB, Cain GD. 1981a. Karyotype evolution and sex chromosome differentiation in schistosomes (Trematoda, Schistosomatidae). Chromosoma (Berl.) 84:413–430.

Grossman AI, Short RB, Kuntz RE. 1981b. Somatic chromosomes of *Schistosoma rodhaini, S. mattheei,* and *S. intercalatum*. Journal of Parasitology 67:41–44.

Gupta BC, Basch PF. 1987a. The role of *Schistosoma mansoni* males in feeding and development of female worms. Journal of Parasitology 73:481–486.

Gupta BC, Basch PF. 1987b. Transfer of a glycoprotein from male to female *Schistosoma mansoni* during pairing. Journal of Parasitology 73:674–675.

Gupta BC, Basch PF. 1989. Human chorionic gonadotropin (hCG)-like immunoreactivity in schistosomes and Fasciola. Parasitology Research 76:86–89.

Haas W. 1988. Host-finding—a physiological effect. Pp. 454–464 In: H. Mehlhorn, Ed. Parasitology in Focus. Berlin, Heidelberg. Springer Verlag.

Haas W, Granzer M, Brockelman CR. 1990. Finding and recognition of the bovine host by the cercariae of Schistosoma spindale. Parasitology Research 76:343–350.

Haas W, Granzer M, Garcia EG. 1987. Host identification by Schistosoma japonicum cercariae. Journal of Parasitology 73:568–577.

Haas W, Schmit R. 1982. Characterization of chemical stimuli for the penetration of Schistosoma mansoni cercariae. 1. Effective substances, host specificity. Zeitschrift für Parasitenkunde 66:293–307.

Hamburger J, Lustigman S, Arap Siongok TK, Ouma JH, Mahmoud AAF. 1982. Characterization of a purified glycoprotein from *Schistosoma mansoni* eggs: Specificity, stability, and the involvement of carbohydrate and peptide moieties in its serological activity. Journal of Immunology 128:1864–1869.

Hang LM, Warren KS, Boros DL. 1974. *Schistosoma mansoni*: antigenic secretions and the etiology of egg granulomas in mice. Experimental Parasitology 35:288–298.

Hansen EL. 1973. Progeny-daughter sporocysts of *Schistosoma mansoni*. International Journal for Parasitology 3:267–268.

Hansen EL, Perez-Mendez G. Long S. Yarwood E. 1973. *Schistosoma mansoni*: Emergence of progeny-daughter sporocysts in monoxenic culture. Experimental Parasitology 33:486–494.

Harn DA, Mitsuyama M, David JR. 1984. *Schistosoma mansoni*: anti-egg monoclonal antibodies protect against cercarial challenge. Journal of Experimental Medicine 159:1371–1387.

Harris AC, Russell RJ, Charters AD. 1984. A review of schistosomiasis in Western Australia, demonstrating the unusual longevity of *Schistosoma mansoni*. Transactions of the Royal Society of Tropical Medicine and Hygiene 78:385–388.

Harris KR. 1975. The fine structure of encapsulation in Biomphalaria glabrata. Annals of the New York Academy of Science 266:446–464.

Haseeb MA, Eveland LK, Fried B. 1985. The uptake, localization and transfer of 4-^{14}C-cholesterol in *Schistosoma mansoni* males and females maintained in vitro. Comparative Biochemistry and Physiology 82A:421–423.

Haseeb MA, Fried B. 1988. Chemical communication in helminths. Advances in Parasitology 27:169–207.

Haseeb MA, Fried B, Eveland LK. 1989. Schistosoma mansoni: female-dependent lipid secretion in males and corresponding changes in lipase activity. International Journal for Parasitology 19:705–710.

Hata H, Orido Y, Yokogawa M, Kojima S. 1988. Schistosoma japonicum and Paragonimus ohirai: antagonism between S. japonicum and P. ohirai in Oncomelania nosophora. Experimental Parasitology 65:125–130.

He Y, Gong Z, Ma J. 1980. Scanning and transmission electron microscopy of *Schistosoma japonicum* egg shell. Chinese Medical Journal 93:861–864.

He Y, Yang H. 1980. Physiological studies on the post-cercarial development of *Schistosoma japonicum*. Acta Zoologica Sinica 26:32–39.

Henkle KJ, Davern KM, Wright MD, Ramos AJ, Mitchell GF. 1990. Comparison of the cloned genes of the 26- and 28-kilodalton glutathione S-transferases of Schistosoma japonicum and Schistosoma mansoni. Molecular and Biochemical Parasitology 40:13–34.

Hess R, Faigle JW, Lambert C. 1966. Selective uptake of an antibilharzial nitrothiazole compound by Schistosoma mansoni. Nature 210:964–965.

Hicks RM, Newman J. 1977. The surface structure of the tegument of Schistosoma haematobium. Cell Biology International Reports 1:157–167.

Higgins-Opitz SB, Evers P. 1983. Observations by scanning electron microscopy of miracidial hatching from Schistosoma mansoni ova. Journal of Parasitology 69:432–433.

Ho Y-h. 1962. Notes on some abnormal specimens of Schistosoma japonicum. Acta Zoologica Sinica 14:453–457.

Ho Y-h, Yang H-c. 1973. Histochemical localization of phenols and phenolase in Schistosoma japonicum. Acta Zoologica Sinica 19:1–10.

Ho Y-h, Yang H-c. 1974. Histological and histochemical studies on the egg formation of Schistosoma japonicum. Acta Zoologica Sinica 20:256–258.

Ho Y-h, Yang H-c. 1979. Histological and histochemical studies on embryonation of ova of Schistosoma japonicum. Acta Zoologica Sinica 25:304–310.

Hockley DJ. 1968. Small spines on the egg shells of Schistosoma. Parasitology 58:367–370.

Hockley DJ. 1973. Ultrastructure of the tegument of Schistosoma. Advances in Parasitology 11:233–305.

Hoeppli R. 1932. Histological observations in experimental schistosomiasis japonica. Chinese Medical Journal 46:1179–1186.

Hoffman DB Jr, Warren KS. 1978. Schistosomiasis IV. Condensations of the Selected Literature 1963–1975. 2 Volumes. Washington DC. Hemisphere Publishing Co.

Howell MJ, Bearup AJ. 1980. The life histories of two bird trematodes of the family Philophthalmidae. Proceedings of the Linnean Society of New South Wales 9:182–194.

Howells RE, Ramalho-Pinto FJ, Gazzinelli G, DeOliveira CC, Figueiredo EA, Pellegrino J. 1974. Schistosoma mansoni: Mechanism of cercarial tail loss and its significance to host penetration. Experimental Parasitology 36:373–385.

Hsü H-c, Hsi F-h, Mao S-p. 1958. Observation on the egg output of Schistosoma japonicum in the mouse host. Chinese Medical Journal 77:559.

Hsü HF, Hsü SYL. 1957. On the intraspecific and interstrain variations of the male sexual glands of Schistosoma japonicum. Journal of Parasitology 43:456–463.

Hsü HF, Hsü SYL. 1958a. On the size and shape of the eggs of the geographic strains of Schistosoma japonicum. American Journal of Tropical Medicine and Hygiene 7:125–134.

Hsü HF, Hsü SYL. 1958b. The prepatent period of four geographic strains of Schistosoma japonicum. Transactions of the Royal Society of Tropical Medicine and Hygiene 52:363–367.

Hsü HF, Hsü SYL. 1960. Susceptibility of albino mice to different geographic strains of Schistosoma japonicum. Transactions of the Royal Society of Tropical Medicine and Hygiene 54:466–468.

Hsü HF, Hsü SYL, Tsai CT. 1960. Further studies on the prepatent periods of four geographic strains Schistosoma japonicum. In: Escuela Nacional de Ciencias

Biologicas, Instituto Politecnico Nacional. Libro Homenaje al Dr. Eduardo Caballero y Caballero. Mexico, D F. p. 153–160.

Hsü SYL, Hsü HF. 1960. On the virulence of the geographic strains of *Schistosoma japonicum*. American Journal of Tropical Medicine and Hygiene 9:195–198.

Hsü SYL, Hsü HF, Chu KY. 1962. Interbreeding of geographic strains of *Schistosoma japonicum*. Transactions of the Royal Society of Tropical Medicine and Hygiene 56:383–385.

Hubendick B. 1958. On the family Ancylidae with special reference to Ferrissia tenuis (Bourguignat), the suspected intermediate host of Schistosoma haematobium in India. Proceedings of the Sixth International Congress on Tropical Medicine and Malaria Vol. 2:17–21.

Hyman LH. 1951. The Invertebrates: Platyhelminthes and Rhynchocoela. The acoelomate Bilateria. Vol. II. New York. The McGraw-Hill Book Co.

Iijima T, Lo CT, Ito Y. 1971. Studies on schistosomiasis in the Mekong basin. 1. Morphological observation of the schistosomes and detection of their reservoir hosts. Japanese Journal of Parasitology 20:24–33.

Imbert-Establet D. 1982. Infestation naturelle des rats sauvages par *Schistosoma mansoni* en Guadeloupe: Données quantitatives sur le développement et la fertilité du parasite. Annales de Parasitologie Humaine et Comparée 57:573–585.

Imbert-Establet D, Rollinson D, Ross G. 1984. Selection de genotypes de *Schistosoma mansoni* et leur maintenance par transplantations sporocystiques. Comptes Rendus de la Académie des Sciences (III) 299:459–462.

Imperia PS, Fried B, Eveland LK. 1980. Pheromonal attraction of *Schistosoma mansoni* females toward males in the absence of worm-like tactile behavior. Journal of Parasitology 66:682–684.

Inatomi S. 1962. Submicroscopic structure of the eggshell of helminth. Okayama Igakkai Zasshi 74(Suppl.):31.

Inatomi S, Sakumoto D, Tongu Y, Sugiuri S, Itano K. 1970. Ultrastructure of *Schistosoma japonicum*. Pp. 257–289 In: M Sasa, M Yokogawa, Eds. Recent advances in Researches on Filariasis and Schistosomiasis in Japan. Tokyo. University of Tokyo Press.

Irie Y, Basch PF, Beach N. 1983. Reproductive ultrastructure of adult Schistosoma mansoni grown in vitro. Journal of Parasitology 69:559–566. [Note: Beach is misprint for Basch]

Irwin SWB, Threadgold LT. 1972. Electron microscope studies of *Fasciola hepatica*. X. Egg formation. Experimental Parasitology 31:321–331.

Isseroff H, Bock K, Owczarek A, Smith KR. 1983. Schistosomiasis: Proline production and release by ova. Journal of Parasitology 69:285–289.

Ito J. 1954. Studies on the host-parasite relationships of *Schistosoma japonicum* in common laboratory animals. Japanese Journal of Medical Science and Biology 8:43–62.

Jackson H, Davies P, Bock M. 1968. Chemosterilization of *Schistosoma mansoni*. Nature 218:977.

Jaffe R, Mayer M, Pifano CF. 1945. Estudios biologicos y anatomo-patologicos en animales infectados con un solo sexo de *Schistosoma mansoni*. Revista de Sanidad y Asistenccia Social 10:95–107.

James ER, Taylor MG. 1976. Transformation of cercariae to schistosomula: A quantitative comparison of transformation techniques and of infectivity by different injection routes of the organisms produced. Journal of Helminthology 50:223–233.

James C, Webbe G. 1973. A comparison of Egyptian and East African strains of *Schistosoma haematobium*. Journal of Helminthology 47:49–59.

James C, Webbe G. 1975. A comparison of Sudanese and South African strains of *Schistosoma haematobium*. Journal of Helminthology 49:191–197.

Jelnes JE. 1983. Phosphoglucose isomerase: sex-linked character in Schistosoma mansoni. Journal of Parasitology 69:780–781.

Jeong KH, Lie KJ, Heyneman D. 1983. The ultrastructure of the amebocyte-producing organ in Biomphalaria glabrata. Developmental and Comparative Immunology 7:217–228.

Johnson KS, Taylor DW, Cordingley JS. 1987. Possible eggshell protein gene from *Schistosoma mansoni*. Molecular and Biochemical Parasitology 22:89–100.

Johri LN, Smyth JD. 1956. A histochemical approach to the study of helminth morphology. Parasitology 46:107–116.

Jones JT, Breeze P, Kusel JR. 1989. Schistosome fecundity: influence of host genotype and intensity of infection. International Journal for Parasitology 19: 769–777.

Jones JT, Kusel JR. 1985. The inheritance of responses to schistosomiasis mansoni in two pairs of inbred strains of mice. Parasitology 90:289–300.

Jones JT, McCaffery DM, Kusel JR. 1983. The influence of the H-2 complex on responses to infection by Schistosoma mansoni in mice. Parasitology 86:19–30.

Jordan P, Webbe G. 1969. Human Schistosomiasis. Springfield, Illinois. Charles C Thomas Publisher.

Jourdane J. 1982. Étude des mécanismes de rejet dans les couples mollusque-schistosome incompatibles à partir d'infestations par voie naturelle et par transplantations microchirurgicales de stades parasitaires. Acta Tropica 39:325–335.

Jourdane J. 1983. Mise en évidence d'un processus original de la reproduction asexuée chez *Schistosoma haematobium*. Comptes Rendus des Séances de l'Académie des Sciences, Paris 296:419–424.

Jourdane J. 1984. Maintenance d'un clone male et d'un clone femelle de *Schistosoma mansoni* par transplantation microchirugicale de sporocystes-fils Viabilité de la methode. Annales de Parasitologie Humaine et Comparée 59:361–367.

Jourdane J, Kechemir N, Combes C. 1981. Mise en évidence d'une replication des sporocystes fils de *Schistosoma haematobium* après transplantation microchirurgicale chez Bulinus truncatus. Comptes Rendus des Séances de l'Académie des Sciences, Paris 293:531–533.

Jourdane J, Liang Y-s, Bruce JI. 1985. Transplantation of *Schistosoma japonicum* daughter sporocysts in Oncomelania hupensis. Journal of Parasitology 71:244–247.

Jourdane J, Mouahid A, Touassem R. 1984. Evolution des sporocystes de *Schistosoma bovis* après transplantation microchirurgicale chez *Bulinus truncatus*. Annales de Parasitologie Humaine et Comparée 59:459–466.

Jourdane J, Théron A. 1980. *Schistosoma mansoni*: Cloning by microsurgical transplantation of sporocysts. Experimental Parasitology 50:349–357.

Jourdane J, Théron A. 1987. Larval development: Eggs to cercariae. In D Rollinson and AJG Simpson, Eds. The Biology of Schistosomes. London. Academic Press.

Jourdane J, Théron A, Combes C. 1980. Demonstration of several sporocysts generations as a normal pattern of reproduction of *Schistosoma mansoni*. Acta Tropica 37:177–182.

Jourdane J, Xia MY. 1987. The primary sporocyst stage in the life cycle of Schistosoma japonicum (Trematoda, Digenea). Transactions of the American Microscopical Society 106:364–372.

Justine JL, Mattei X. 1981. Étude ultrastructurale du flagelle spermatique des schistosomes (Trematoda: Digenea). Journal of Ultrastructure Research 67:89–95.

Kamiya H, Smithers SR, McLaren DJ. 1987. Schistosoma mansoni: autoradiographic tracking studies of isotopically-labelled challenge parasites in naive and vaccinated CBA/Ca mice. Parasite Immunology 9:515–529.

Kassim OO, Cheever AW, Richards CS. 1979. Schistosoma mansoni: mice infected with different worm strains. Experimental Parasitology 48:220–224.

Kassim OO, Richards CS. 1979. Radioisotope labeling of Schistosoma mansoni miracidia for in vivo studies in Biomphalaria glabrata. Journal of Invertebrate Pathology 33:385–386.

Katsumata T, Kohno S, Yamaguchi K, Hara K, Aoki Y. 1989. Hatching of Schistosoma mansoni eggs is a Ca^{2+}/calmodulin-dependent process. Parasitology 76:90–91.

Katsumata T, Shimada M, Sato K, Aoki Y. 1988. Possible involvement of calcium ions in the hatching of Schistosoma mansoni eggs in water. Journal of Parasitology 74:1040–1041.

Katsurada, F. 1904. Schistosomum japonicum, ein neuer menschlichen Parasit, durch welchen eine endemische Krankheit in verschiedenen Gegended Japans verursacht wird. Annotationes Zoologicae Japonense 5:147–160.

Kawamoto F, Shozawa A, Kumada N, Kojima K. 1989. Possible roles of cAMP and Ca^{2+} in the regulation of miracidial transformation in Schistosoma mansoni. Parasitology Research 75:368–374.

Kechemir N, Combes C. 1982. Développement du trematode Schistosoma haematobium après transplantation microchirurgicale chez le gasteropode Planorbarius metidjensis. Comptes Rendus des Séances l'Académie des Sciences, Paris 295:505–508.

Kechemir N, Théron A. 1980. Existence of replicating sporocysts in the development cycle of Schistosoma haematobium. Journal of Parasitology 66:1068–1080.

Kee KC, Taylor DW, Cordingey JS, Butterworth AE, Munro AJ. 1986. Genetic influence on the antibody response to antigens of Schistosoma mansoni in chronically infected mice. Parasite Immunology 8:565–574.

Kelley DH, von Lichtenberg F. 1970. "Abnormal" schistosome oviposition. Origin of aberrant shell structures and their appearance in human tissues. American Journal of Pathology 60:271–287.

Khayyal MT. 1964. The effects of antimony uptake on the location and pairing of Schistosoma mansoni. British Journal of Pharmacology 22:342–348.

Khayyal MT, Kheir Eldin AA, Saleh S, Sayed Ahmed HM, Zein El-Abedin A, Roshdi MZ, Metwally AA. 1980. Effect of oxamniquine on the lipid pattern of female and male S. mansoni worms. Journal of The Egyptian Medical Association 63:195–204.

Kinoti GK. 1971. The attachment apparatus and penetration apparatus of the miracidium of Schistosoma. Journal of Helminthology 45:229–235.

Kinoti GK. 1987. The significance of variation in the susceptibility of Schistosoma mansoni to the antischistosomal drug Oxaminiquine. Memorias do Instituto Oswaldo Cruz 82 Suppl 4:151–156.

Kitajima EW, Paraense WL, Corrêa LR. 1976. The fine structure of Schistosoma mansoni sperm (Trematoda: Digenea). Journal of Parasitology 2:215–221.

Kitikoon V. 1980. Comparison of eggs and miracidia of *Schistosoma mekongi* and *S. japonicum*. Malacological Review Suppl. 2:83–103.

Kloetzel K. 1967a. A collagenase-like enzyme diffusing from eggs of *Schistosoma mansoni* (Correspondence). Transactions of the Royal Society of Tropical Medicine and Hygiene 61:608–609.

Kloetzel K. 1967b. Egg and pigment production in *Schistosoma mansoni* infections of the white mouse. American Journal of Tropical Medicine and Hygiene 16:293–299.

Kloetzel K. 1968. A collagenaselike substance produced by eggs of Schistosoma mansoni. Journal of Parasitology 54:177–178.

Knauft RF, Warren KS. 1969. The effect of calorie and protein malnutrition on both the parasite and the host in acute urine schistosomiasis mansoni. Journal of Infectious Diseases 120:560–575.

Knight M, Simpson AJG, Kelly C, Smithers R. 1986. The cloning of schistosome genes encoding antigenic and maturation-like polypeptides. Parasitology 91 Suppl:S39–S51.

Knopf PM. 1982. The role of host hormones in controlling survival and development of *Schistosoma mansoni*. Pharacology and Therapeutics 15:293–311.

Knopf PM, Linden T. 1985. Completion of *Schistosoma mansoni* life cycle in thyroidectomized rats and effects of thyroid hormone replacement therapy. Journal of Parasitology 71:422–426.

Køie M, Frandsen F. 1976. Stereoscan observations of the miracidium and early sporocyst of Schistosoma mansoni. Zeitschrift für Parasitenkunde 50:335–344.

Koura M. 1970. The relation between egg production and worm burden in experimental schistosomiasis. I. Worm burden. Journal of The Egyptian Medical Association 53:589–602.

Koura M. 1971a. The relation between egg production and worm burden in experimental schistosomiasis. III. Effect of repeated exposure to infection. Journal of The Egyptian Medical Association 54:709–713.

Koura M. 1971b. The relation between egg production and worm burden in schistosomiasis. II. Duration of infection. Journal of the Egyptian Medical Association 55:44–450.

Krakower C, Hoffman WA, Axtmayer JH. 1940. The fate of schistosomes (*S. mansoni*) in experimental infection of normal and vitamin A deficient white rats. Puerto Rican Journal of Tropical Medicine and Public Health 16:269–304.

Krakower C, Hoffman WA, Axtmayer JH. 1944. Defective granular eggshell formation by *Schistosoma mansoni* in experimentally infected guinea pigs on a vitamin C deficient diet. Journal of Infectious Diseases 74:178–183.

Kruatrachue M, Bhaibulaya M, Harinasuta C. 1965. Orientobilharzia harinasutai sp. nov., a mammalian blood-fluke, its morphology and life cycle. Annals of Tropical Medicine and Parasitology 59:181–188.

Kruatrachue M, Riengrojpitak S, Upatham ES, Sahaphong S. 1983a. Scanning electron microscopy of the tegumental surface of adult *Schistosoma spindale*. Southeast Asian Journal of Tropical Medicine and Publich Health 14:281–289.

Kruatrachue M, Upatham ES, Sahaphong S, Riengrojpitak S. 1979. Scanning and transmission electron microscopy of Mekong Schistosoma eggs and adults. Southeast Asian Journal of Tropical Medicine and Public Health 14:85–96.

Kruatrachue M, Upatham ES, Sahaphong S, Tongtong T, Khunborivan V. 1983b. Scanning electron microscopic study of the tegumental surface of adult *Schis-*

tosoma sinensium. Southeast Asian Journal of Tropical Medicine and Public Health 14:427–438.

Kruger FJ, Hamilton-Atwell VL, Schutte CH. 1986a. Scanning electron microscopy of the tegument of males from five populations of *Schistosoma mattheei.* Onderstepoort Journal of Veterinary Science 53:109–110.

Kruger FJ, Schutte CH, Visser PS, Evans AC. 1986b. Phenotypic differences in *Schistosoma mattheei* ova from populations sympatric and allopatric to *Schistosoma haematobium.* Onderstepoort Journal of Veterinary Science 53:103–107.

Kruger SP, Heitman LP, van Wyk JA, McCully RM. 1969. The route of migration of *Schistosoma mattheei* from the lungs to the liver in sheep. Journal of the South African Veterinary Medical Association 40:39–43.

Krupa PL, Lewis LM, del Vechio P. 1977. Schistosoma haematobium in Bulinus guernei: Electron microscopy of hemocyte-sporocyst interactions. Journal of Invertebrate Pathology 30:35–45.

Kuntz RE. 1955. Biology of the schistosome complexes. American Journal of Tropical Medicine and Hygiene 4:383–414.

Kuntz RE. 1961a. Passage of eggs by hosts infected with *Schistosoma haematobium.* Journal of Helminthology R T Leiper Suppl:107–116.

Kuntz RE. 1961b. Passage of eggs by hosts infected with *Schistosoma mansoni* with emphasis on rodents. Journal of Parasitology 47:905–909.

Kuntz RE, Davidson DL, Huang TC, Tulloch GS. 1979. Scanning electron microscopy of the integumental surfaces of *Schistosoma bovis.* Journal of Helminthology 53:131–132.

Kuntz RE, Tulloch GS, Davidson DL, Huang T-c. 1976. Scanning electron microscopy of the integumental surfaces of *Schistosoma haematobium.* Journal of Parasitology 62:63–69.

Kunz W, Symmons P. 1987. Gender-specifically expressed genes in Schistosoma mansoni. Acta Tropica 44 Suppl 12:90–93.

Kusel JR. 1970. Studies on the structure and hatching of the eggs of *Schistosoma mansoni.* Parasitology 60:79–88.

Lagrange E. 1954. Le sexe des cercaires de *Schistosoma mansoni.* Rivista di Parassitologia 25:81–84.

Lagrange E. 1958. Infections unisexuées et possibilité de guérison de bilharziose à S. mansoni chez la souris. Rivista di Parassitologia 19:59–66.

Lagrange E, Scheecqmans G. 1949. La bilharziose expérimentale du cobaye. Comptes Rendus des Séances de la Société de Biologie et de ses Filiales (Paris), Paris 143:1396–1399.

Lancastre F, Coutris G, Bolognini-Treney J, Traore L, Mougeot G. 1984. Schistosomose experimentale III. Observations sur l'infestation de *Biomphalaria glabrata* par un miracidium de *Schistosoma mansoni.* Annales de Parasitologie Humaine et Comparée 59:79–94.

Lawrence JA. 1973a. *Schistosoma mattheei* in cattle: The host-parasite relationship. Research in Veterinary Science 14:400–402.

Lawrence JA. 1973b. *Schistosoma mattheei* in cattle: Variations in parasite egg production. Research in Veterinary Science 14:402–404.

Lawrence JD. 1973. The ingestion of red blood cells by *Schistosoma mansoni.* Journal of Parasitology 59:60–63.

Lawson JR, Wilson RA. 1980. Metabolic changes associated with the migration of the schistosomulum of *Schistosoma mansoni* in the mammal host. Parasitology 82:325–326.

Lawson R, Draskau T. 1977. The development of the schistosomulum of *Schistosoma mansoni* during its migration in the mammalian host. Transactions of the Royal Society of Tropical Medicine and Hygiene 71:289.

Lee H-F. 1962. Life history of Heterobilharzia americana Price 1929, a schistosome of the raccoon and other mammals in southeastern United States. Journal of Parasitology 48:728–739.

Lee HG. 1972. Aspects of the effect of thioxanthone on *Schistosoma mansoni* in mice and in vitro. Bulletin of the World Health Organization 46:397–402.

Leiper RT. 1916. On the relation between the terminal-spined and lateral-spined eggs of Bilharzia. British Medical Journal 1:411.

Leiper RT. 1918. Report on the results of the Bilharzia mission in Egypt, 1915. Part V.—Adults and ova. Journal of the Royal Army Medical Corps 30:235–253.

Leitch B, Probert AJ, Runham NW. 1984. The ultrastructure of the tegument of adult *Schistosoma haematobium*. Parasitology 89:71–78.

Lengy J. 1962a. Studies on *Schistosoma bovis* (Sonsino, 1876) in Israel I. Larval stages from egg to cercaria. Bulletin of the Research Council of Israel 10E:1–36.

Lengy J. 1962b. Studies on *Schistosoma bovis* (Sonsino, 1876) in Israel. II. The intra-mammalian phase of the life cycle. Bulletin of the Research Council of Israel 10E:73–96.

Lennox RW, Schiller EL. 1972. Changes in the dry weight and glycogen content as criteria for measuring the postcercarial growth and development of *Schistosoma mansoni*. Journal of Parasitology 58:489–494.

LeRoux PL. 1949. Abnormal adaptations and development of schistosomes in experimental animals. Transactions of the Royal Society of Tropical Medicine and Hygiene 43:2–3.

LeRoux PL. 1951a. Abnormalities in females of *Schistosoma mattheei*. Transactions of the Royal Society of Tropical Medicine and Hygiene 44:364.

LeRoux PL. 1951b. Hermaphroditism in males of *Schistosoma mansoni* carrying females in the gynaecophoric canal. Transactions of the Royal Society of Tropical Medicine and Hygiene 44:363.

LeRoux PL. 1954. Hybridisation of *Schistosoma mansoni* and *Schistosoma rodhaini*. Transactions of the Royal Society of Tropical Medicine and Hygiene 48:3–4.

LeRoux PL. 1955. A new mammalian schistosome (Schistosoma leiperi sp. nov.) from herbivora in Northern Rhodesia. Transactions of the Royal Society of Tropical Medicine and Hygiene 49:293–294.

Li P-c. 1958. Development of *Schistosoma japonicum* in unisexual and bisexual infections. Chinese Medical Journal 77:559.

Liberatos JD, Short RB. 1983. Identification of sex of schistosome larval stages. Journal of Parasitology 69:1084–1089.

Liberatos JD. 1987. Schistosoma mansoni: male-biased sex ratios in snails and mice. Experimental Parasitology 64:165–177.

Liberti P, Festucci A, Ruppel A, Gigante S, Cioli D. 1986. A surface-labeled 18 kilodalton antigen in Schistosoma mansoni Molecular and Biochemical Parasitology 18:55–67.

Lichtenberg F, Lindenberg M. 1957. An alcohol-acid-fast substance in eggs of *Schis-*

tosoma mansoni. American Journal of Tropical Medicine and Hygiene 3:1066–1076.

Lie KJ. 1973. Larval trematode antagonism principles and possible application as control method. Experimental Parasitology 33:343–349.

Lie KJ, Basch PF, Heyneman D. 1968a. Direct and indirect antagonism between Paryphostomum segregatum and Echinostoma paraensei in the snail Biomphalaria glabrata. Zeitschrift für Parasitenkunde 31:101–107.

Lie KJ, Basch PF, Heyneman D, Beck AJ, Audy JR. 1986b. Implications for trematode control of interspecific larval antagonism within snail hosts. Transactions of the Royal Society of Tropical Medicine and Hygiene 62:299–319.

Lie KJ, Basch PF, Umathevy T. 1965. Antagonism between two species of larval trematodes in the same snail. Nature 206:422–423.

Lie KJ, Heyneman D, Yau P. 1975. The origin of amebocytes in Biomphalaria glabrata. Journal of Parasitology 61:574–576.

Lim HK, Heyneman D. 1972. Intermolluscan inter-trematode antagonism: A review of factors influencing the host-parasite system and its possible role in biological control. Advances in Parasitology 10:191–268.

Lindner E. 1914. Über die Spermatogenese von Schistosomum haematobium Bilh. (Bilharzia haematobia Cobb.) mit besonderer Berücksichtungen der Geschlechtschromosomen. Archiv für Zellforschung 12:516–538.

Liston WG, Soparkar MB. 1918. Bilharziasis among animals in India. The life cycle of *Schistosomum spindalis.* Indian Journal of Medical Research 5:567–569.

Loison G, Vidal A, Findeli A, Roitsch C, Balloul JM, Lemoine Y. 1989. High level of expression of a protective antigen of schistosomes in Saccharomyces cerevisiae. Yeast 5:497–507.

Loker ES, Bayne CJ, Buckley PM, Kruse KT. 1982. Ultrastructure of encapsulation of Schistosoma mansoni mother sporocysts by hemocytes of juveniles of the 10-R2 strain of Biomphalaria glabrata. Journal of Parasitology 68:84–94.

Loker ES. 1983. A comparative study of the life-histories of mammalian schistosomes. Parasitology 87:343–369.

Looss A. 1895. Zur Anatomie und Histologie von Bilharzia haematobia (Cobbold). Archiv für Mikroskopische Anatomie 46:1–108.

LoVerde PT. 1976. Scanning electron microscopy of the ova of *Schistosoma haematobium* and *Schistosoma mansoni.* Egyptian Journal of Bilharziasis 3:69–72.

LoVerde PT, DeWald J, Minchella DJ. 1985a. Further studies on genetic variation in *Schistosoma mansoni.* Journal of Parasitology 71:732–734.

LoVerde PT, DeWald J, Minchella DJ, Bosshardt SC, Damian RT. 1985b. Evidence for host-induced selection in *Schistosoma mansoni.* Journal of Parasitology 71:297–301.

LoVerde PT, Rekosh DM, Bobek LA. 1989. Developmentally regulated gene expression in Schistosoma. Experimental Parasitology 68:116–120.

Lutz A. 1919. Schistosomum Mansoni and schistosomiasis observed in Brazil. Memorias do Instituto Oswaldo Cruz 11:109–140 + pls.

Lwambo NJ, Upatham ES, Kruatrachue M, Viyanant V. 1987. The host-parasite relationship between the Saudi Arabian Schistosoma mansoni and its intermediate and definitive hosts. 2. Effects of temperature, salinity and pH on the infection of mice by Schistosoma mansoni cercariae. Southeast Asian Journal of Tropical Medicine and Public Health 18:166–170.

Ma J, He Y. 1981. Scanning electron microscopy of Chinese (Mainland) strain of *Schistosoma japonicum.* Chinese Medical Journal 94:63–70.

Machado ABM, Henriques da Silva, ML, Pellegrino, J. 1970. Mechanism of action of thiosinamine, an egg suppressive agent in schistosomiasis. Journal of Parasitology 56:392–393.

Machattie C. 1936. A preliminary note on the life history of *Schistosoma turkestanicum* Skrjabin, 1913. Transactions of the Royal Society of Tropical Medicine and Hygiene 30:115–124.

Maciel H. 1924. Algumas notas sobre a postura do "*Schistosoma mansoni*". Sciencia Medica (Rio) 2:419–425.

Maciel H. 1926. A influencia des forças cosmicas sobre a postura do *Schistosoma mansoni*. Sciencia Medica (Rio) 4:589–594.

Macy RW, Basch PF. 1972. Orthetrotrema monostomum gen. et sp. n., a progenetic trematode (Dicrocoeliidae) from dragonflies in Malaysia. Journal of Parasitology 58:515–518.

Magalhães LA, de Carvalho JF. 1973. Estudo morfologico de *Schistosoma mansoni* pertenecentes a linhagens de Belo Horizonte (MG) e de Sao Jose dos Campos (SP). Revista de Saude Publica, Sao Paulo 7:289–294.

Magath TB, Mathieson DR. 1946. Factors affecting the hatching of ova of *Schistosoma japonicum*. Journal of Parasitology 32:64–68.

Maldonado JF, Herrera FV. 1950. *Schistosoma mansoni* infection resulting from exposure to cercariae proceeding from single, naturally infected snails. Puerto Rico Journal of Public Health and Tropical Medicine 25:230–241.

Malek EA. 1961. The biology of mammalian and bird schistosomes. Bulletin of the Tulane University Medical Faculty 20:181–207.

Malek EA. 1985. Snail Hosts of Schistosomiasis and Other Snail-Transmitted Diseases in Tropical America: A Manual. Washington, DC. Pan American Health Organization Scientific Publication 478:1–325.

Manson-Bahr P, Fairley NH. 1920. Observations on bilharziasis amongst the Egyptian expeditionary force. Parasitology 12:33–71.

Mao C-P, Li L. 1948. A note on the morphology of *Schistosoma japonicum*. Proceedings of the Helminthological Society of Washington 15:77–79.

Marikovsky M, Fishelson Z, Arnon R. 1988. Purification and characterization of proteases secreted by transforming schistosomula of Schistosoma mansoni. Molecular and Biochemical Parasitology 30:45–54.

Markel SF, LoVerde PT, Britt EM. 1978. Prolonged latent schistosomiasis. Journal of the American Medical Association 240:1746–1747.

Massoud J. 1973a. Studies on the schistosomes of domestic animals in Iran: I. Observations on Ornithobilharzia turkestanicum (Skrjabin, 1913) in Khuzestan. Journal of Helminthology 47:165–180.

Massoud J. 1973b. Parasitological and pathological observations on *Schistosoma bovis* Sonsino, 1876, in calves, sheep and goats in Iran. Journal of Helminthology 47:155–164.

Matsuda H, Tanaka H, Nogami S, Muto M. 1983. Mechanism of action of praziquantel on the egg of Schistosoma japonicum. Japanese Journal of Experimental Medicine 53:271–274.

Mayer M, Pifano CF, 1942. Estudios biologicos y patologicos en animales infectados con *Schistosoma mansoni* (infecciones bi- y unisexuales). Revista de Sanidad y Asistencia Social 7:419–428.

McCully RM, Kruger SP. 1969. Observations on bilharziasis of domestic ruminants in South Africa. Onderstepoort Journal of Veterinary Research 36:129–162.

McCutchan TF, Simpson AJG, Mullins J, Sher A, Nash T, Lewis F, Richards C.

1984. Differentiation of schistosomes by species, strain, and sex using cloned DNA markers. Proceedings of the National Academy of Sciences. U.S.A. 81:889–893.

McKerrow JH, Doenhoff MJ. 1988. Schistosome proteases. Parasitology Today 4:334–340.

McKerrow JH, Jones P, Sage H, Pino-Heiss S. 1985. Proteinases from invasive larvae of the trematode parasite Schistosoma mansoni degrade connective-tissue and basement-membrane macromolecules. Biochemical Journal 231:47–51.

McLaren DJ. 1980. Schistosoma mansoni: The Parasite Surface in Relation to Host Immunity. Chichester (UK). Research Studies Press.

McLaren DJ, Smithers SR. 1987. The immune response to schistosomiasis in experimental hosts. Pp. 233–263 In: D Rollinson and AJG Simpson, Eds. The Biology of Schistosomes From Genes to Latrines. London. Academic Press.

McMullen DB, Beaver PC. 1945. Studies on schistosome dermatitis. IX. The life cycles of three dermatitis-producing schistosomes from birds and a discussion of the subfamily Bilharziellinae (Trematoda: Schistosomatidae). American Journal of Hygiene 42:128–154.

Medley G, Anderson RM. 1985. Density-dependent fecundity in Schistosoma mansoni infections in man. Transactions of the Royal Society for Tropical Medicine and Hygiene 79:532–534.

Mellink JJ, van den Bovenkamp WW. 1985. In vitro culture of intramolluscan stages of the avian schistosome Trichobilharzia ocellata. Zeitschrift für Parasitenkunde 71:337–351.

Meuleman EA. 1972. Host-parasite interrelationships between the freshwater pulmonate Biomphalaria pfeifferi and the trematode Schistosoma mansoni. Netherlands Journal of Zoology 22:355–427.

Meuleman EA, Holzmann PJ, Peet RC. 1980. The development of daughter sporocysts inside the mother sporocyst of schistosoma mansoni; with special reference to the ultrastructure of the body wall. Zeitschrift für Parasitenkunde 61:201–212.

Meuleman EA, Lyaruu DM, Khan MA, Holzmann PJ, Sminia T. 1978. Ultrastructural changes in the body wall of Schistosoma mansoni during the transformation of the miracidium into the mother sporocyst in the snail host Biomphalaria glabrata. Zeitschrift für Parasitenkunde 56:227–242.

Meuleman EA, te Velde AA. 1984. Age-dependent susceptibility of Lymnaea stagnalis to Trichobilharzia ocellata. Tropical and Geographic Medicine 36:387.

Meusel G, Symmons P, Kunz W. 1987. Two gender-specific transcripts predominate in the mRNA of adult female Schistosoma mansoni. Journal of Parasitology 73:436–437.

Michaels RM. 1969. Mating of Schistosoma mansoni in vitro. Experimental Parasitology 25:58–71.

Michaels RM. 1970. S. mansoni: alteration of ovipositing capacity by transplanting between heterologous hosts. Experimental Parasitology 27:217–228.

Michaels RM, Prata A. 1968. Evolution and characteristics of Schistosoma mansoni eggs laid in vitro. Journal of Parasitology 54:921–930.

Michelson EH. 1986. Preliminary observations on a hemolymph factor influencing the infectivity of Schistosoma mansoni miracidia. Acta Tropica 43:63–68.

Michelson EH, DuBois L. 1981. Resistance to schistosome infection in Biomphalaria glabrata induced by gamma radiation. Journal of Invertebrate Pathology 38:39–43.

Miller P, Wilson RA. 1980. Migration of the schistosomula of *Schistosoma mansoni* from the lungs to the hepatic portal system. Parasitology 80:267–288.

Mitchell GF. 1989. Portal system peculiarities may contribute to resistance against schistosomes in 129/J mice. Parasite Immunology 11:713–717.

Miyagawa Y, Takemoto S. 1921. The mode of infection of *Schistosomum japonicum* and the principal route of its journey from the skin to the portal vein in the host. Journal of Pathology and Bacteriology 24:168–174.

Moczon T, Swiderski Z. 1983. Schistosoma haematobium: Oxidoreductase histochemistry and ultrastructure of niridazole-treated females. International Journal for Parasitology 13:225–232.

Mohandas A. 1983. 4. Platyhelminthes—Trematoda. Pp. 105–129 In: KG Adiyodi, RG Adiyodi, Eds. Reproductive Biology of Invertebrates. Vol. II. Chichester. John Wiley & Sons.

Molokhia MM, Smith H. 1968. Antimony uptake by schistosomes. Annals of Tropical Medicine and Parasitology 62:158–163.

Moloney NA, Webbe G. 1983. The host-parasite relationship of *Schistosoma japonicum* in CBA mice. Parasitology 87:327–342.

Monteiro W, Pellegrino J, Da Silva MLH. 1949. Some physiological and morphological changes in egg formation of *Schistosoma mansoni* following treatment with antischistosomal agents. Revista Brasileira de Pesquisas Medicas e Biologicas 2:45.

Moore DV. 1955. Sexual anomalies in the male *Schistosoma japonicum* (Formosa strain). Journal of Parasitology 41:104–108.

Moore DV, Meleney HE. 1954. Comparative susceptibility of common laboratory animals to experimental infection with *Schistosoma haematobium*. Journal of Parasitology 40:392–397.

Moore DV, Sandground JH. 1956. The relative egg producing capacity of *Schistosoma mansoni* and *Schistosoma japonicum*. American Journal of Tropical Medicine and Hygiene 5:831–840.

Moore DV, Yolles TK, Meleney HE. 1949. A comparison of common laboratory animals as experimental hosts for *Schistosoma mansoni*. Journal of Parasitology 35:156–170.

Moore DV, Yolles TK, Meleney HE. 1954. The relationship of male worms to the sexual development of female *Schistosoma mansoni*. Journal of Parasitology 40:166–185.

Morris GP. 1968. Fine structure of the gut epithelium of *Schistosoma mansoni*. Experientia 24:480–482.

Moulchet F, Develoux M, Magasa MB. 1988. Schistosoma bovis in human stools in Republic of Niger. Transactions of the Royal Society of Tropical Medicine and Hygiene 82:257.

Muller RL, Taylor MG. 1972. On the use of the Ziehl-Neelsen technique for specific identification of schistosome eggs. Journal of Helminthology 46:139–142.

Mulvey M, Woodruff DS. 1985. Genetics of *Biomphalaria glabrata*: linkage analysis of genes for pigmentation, enzymes, and resistance to *Schistosoma mansoni*. Biochemical Genetics 23:877–889.

Mutani A, Christensen NO, Frandsen F. 1985. A study of the biological characteristics of a hybrid line between male *Schistosoma haematobium* (Dar es Salaam, Tanzania) and female *S. intercalatum* (Edea, Cameroun). Acta Tropica 42:319–331.

Nagai Y, Gazzinelli G, de Moraes GWG, Pellegrino J. 1977. Protein synthesis during cercaria-schistosomula transformation and early development of the Schistosoma mansoni larvae. Comparative Biochemistry and Physiology 57B: 27–30.

Najim AT. 1951. A male *Schistosoma mansoni* with two sets of testes. Journal of Parasitology 37:545–546.

Nara T, Iwamura Y, Tanaka M, Irie Y, Yasuraoka K. 1990. Dynamic changes of DNA sequences in Schistosoma mansoni in the course of development. Parasitology 100:241–245.

Neill PJ, Smith JH, Doughty BL, Kemp M. 1988. The ultrastructure of the Schistosoma mansoni egg. American Journal of Tropical Medicine and Hygiene 39:52–65.

Nelson GS, Saoud MFA. 1966. The daily egg output of *Schistosoma mansoni* in rhesus monkeys. Transactions of the Royal Society of Tropical Medicine and Hygiene 60:429–430.

Nelson GS, Saoud MFA. 1968. A comparison of the pathogenicity of two geographical strains of *Schistosoma mansoni* in rhesus monkeys. Journal of Helminthology 42:339–362.

Nene V, Dunne DW, Johnson KS, Taylor DW, Cordingley JS. 1986. Sequence and expression of a major antigen from *Schistosoma mansoni*. Homologies to heat shock proteins and alpha-crystallins. Molecular and Biochemical Parasitology 21:179–188.

Newport GR, Weller TH. 1982. Deposition and maturation of eggs of *Schistosoma mansoni* in vitro: importance of fatty acids in serum-free media. American Journal of Tropical Medicine and Hygiene 31:349–357.

Newsome J. 1963. Observations on corticosteroid treatment of schistosomiasis in hamsters and baboons. Transactions of the Royal Society of Tropical Medicine and Hygiene 57:425–432.

Newton WL. 1952. The comparative tissue reaction of two strains of Australorbis glabratus to infection with Schistosoma mansoni. Journal of Parasitology 38:362–366.

Newton WL. 1593. The inheritance of susceptibility to infection with Schistosoma mansoni in Australorbis glabratus. Experimental Parasitology 2:242–257.

Nez MM, Short RB. 1957. Gametogenesis in *Schistosomatium douthitti* (Cort) (Schistosmatidae: Trematoda). Journal of Parasitology 43:167–182.

Niemann GM, Lewis FA. 1990. Schistosoma mansoni: influence of Biomphalaria glabrata size on susceptibility to infection and resultant cercarial production. Experimental Parasitology 70:286–292.

Nirde P, DeReggi ML, Tsoupras G, Torpier G, Fressancourt P, Capron A. 1984. Excretion of ecdysteroids by schistosomes as a marker of parasite infection. FEBS Letters 168:235–240.

Nirde P, Torpier G, DeReggi ML, Capron A. 1983. Ecdysone and 20 hydroxyecdysone: new hormones for the human parasite *Schistosoma mansoni*. FEBS letters 151:223–227.

Niyamasena SG. 1940. Chromosomen und Geschlecht bei Bilharzia mansoni. Zeitschrift für Parasitenkunde 11, 690–710.

Nollen PM. 1971. Digenetic trematodes: Quinone tanning system in eggshells. Experimental Parasitology 30:64–67.

Nollen PM. 1975. The movement and development of reproductive cells within the reproductive systems of *Schistosoma mansoni, S. japonicum*, and *S. haematobium*. Abstracts, U.S.-Japan Cooperative Medical Science Program

Joint Conference on Parasitic Diseases, Bethesda, Maryland Oct 27–29. p. 107–111.

Nollen PM. 1981. Localization of 3H-thymidine in oocyte mitochondria from *Schistosoma mansoni*. Journal of Parasitology 67:355–361.

Nollen PM. 1983. Patterns of sexual reproduction among parasitic platyhelminths. Parasitology 86:99–120.

Nollen PM, Floyd RD, Kozlow RG, Deters DL. 1976. The timing of reproductive cell development and movement in *Schistosoma mansoni, S. japonicum,* and *S. haematobium,* using techniques of autoradiography and transplantation. Journal of Parasitology 62:227–231.

Norden A, Strand M. 1984. *Schistosoma mansoni, S. haematobium,* and *S. japonicum:* identification of genus-, species-, and gender-specific antigenic worm glycoproteins. Experimental Parasitology 57:110–123.

Nuttman CJ. 1971. The fine structure of ciliated nerve endings in the cercaria of Schistosoma mansoni. Journal of Parasitology 57:855–859.

Obasi OE. 1986. Cutaneous schistosomiasis in Nigeria: an update. British Journal of Dermatology 114:597–602.

Ogbe MG. 1982. Scanning electron microscopy of tegumental surfaces and developing *Schistosoma margrebowiei* LeRoux, 1933. International Journal for Parasitology 12:191–198.

Ogbe MG. 1983. In vivo and in vitro development of *Schistosoma margrebowiei*. Journal of Helminthology 57:231–235.

Ohta H, Hayashi M, Tormis LC, Blas BC, Nosenas JS, Sasazuki T. 1987. Immunogenetic factors involved in the pathogenesis of distinct clinical manifestations of schistosomiasis japonica in the Philippine population. Transactions of the Royal Society of Tropical Medicine and Hygiene 81:292–296.

Olivier LJ. 1966. Infectivity of Schistosoma mansoni cercariae. American Journal of Tropical Medicine and Hygiene 15:882–885.

Onori E. 1962. Observations on variations in Schistosoma haematobium egg output, and on the relationship between the average egg output of infected persons and the prevalence of infection in a community. Annals of Tropical Medicine and Parasitology 56:292–296.

Orido Y. 1989. Histochemical evidence of the catecholamine-associated nervous system in certain schistosome cercariae. Parasitology Research 76:146–149.

Ottolina C. 1957. El miracidio del Schistosoma mansoni; anatomia—citologia—fisiologia. Revista de Sanidad y Asistencia Social (Venezuela) 22:1–411.

Otubanjo OA. 1980. *Schistosoma mansoni:* the ultrastructure of the ducts of the male reproductive system. Parasitology 81:565–571.

Otubanjo OA. 1981a. *Schistosoma mansoni:* Astiban-induced damage to tegument and the male reproductive system. Experimental Parasitology 52:161–170.

Otubanjo OA. 1981b. *Schistosoma mansoni:* The sustentacular cells of the testes. Parasitology 82:125–130.

Page MR, Huizinga HW. 1976. Intramolluscan trematode antagonism in a genetically susceptible strain of Biomphalaria glabrata. International Journal for Parasitology 6:117–120.

Pan SC-T. 1980. The fine structure of the miracidium of *Schistosoma mansoni*. Journal of Invertebrate Pathology 36:307–372.

Paperna I. 1967. The effect of pre-existing trematode infection on the establishment of Schistosoma haematobium larvae in bulinid snails. Ghana Medical Journal 6:89–90.

Paperna I. 1968. Susceptibility of Bulinus (Physopsis) globosus and Bulinus truncatus rohlfsi from different localities in Ghana to different local strains of *Schistosoma haematobium*. Annals of Tropical Medicine and Parasitology 62:13–26.

Paraense WL. 1949. Further observations on the sex of *Schistosoma mansoni* in the
infestations produced by cercariae from a single snail. Memorias do Instituto
Oswaldo Cruz 47:547–556.

Paraense WL, Malheiros Santos J. 1949. The sex of *Schistosoma mansoni* in the
infestations produced by cercariae from a single snail. Memorias do Instituto
Oswaldo Cruz 47:51–62.

Pellegrino J, Coelho PM. 1978. *Schistosoma mansoni*: wandering capacity of a worm
couple. Journal of Parasitology 64:181–182.

Pellegrino J, Faria J. 1965. The oogram method for the screening of drugs in *Schistosomiasis mansoni*. American Journal of Tropical Medicine and Hygiene
14:363–369.

Pesigan TP, Farooq M, Hairston NG, Jauregui JJ, Garcia EG, Santos AT, Santos
BC, Besa AA. 1958. Studies of *Schistosoma japonicum* infection in the Phil-
ippines. 1. General considerations and epidemiology. Bulletin of the World
Health Organization 18:345–455.

Pica Mattoccia L, Lelli A, Cioli D. 1982. Sex and drugs in *Schistosoma mansoni*.
Journal of Parasitology 68:347–349.

Pifano CF. 1948. La infeccion uni-sexual producida por *Schistosoma mansoni* en
condiciones experimentales. Archivos Venezolanos de Patologia Tropical y
Parasitologia 1:63–72.

Pitchford RJ. 1959. Cattle schistosomiasis in man in the eastern Transvaal. Transactions of the Royal Society of Tropical Medicine and Hygiene 53:285–290.

Pitchford RJ. 1961. Observations on a possible hybrid between the two schistosomes
S. haematobium and *S. mattheei*. Transactions of the Royal Society of Tropical Medicine and Hygiene 55:44–51.

Pitchford RJ. 1965. Differences in the egg morphology and certain biological characteristics of some African and middle Eastern schistosomes, genus Schistosoma, with terminal-spined eggs. Bulletin of the World Health Organization
32:105–120.

Pitchford RJ. 1977. A check list of definitive hosts exhibiting evidence of the genus
Schistosoma Weinland, 1858 acquired naturally in Africa and the Middle East.
Journal of Helminthology 51:229–252.

Piva N, deCarneri S. 1961. Studio istochimico sui vitellogeni di *Schistosoma mansoni*. Parassitologia 3:235–238.

Piva N, Grimley PM. 1966. Studies on the secretion of the vitellogenous gland of
Schistosoma mansoni. Journal of Cell Biology 31:155A.

Platt TR, Blair D, Purdie J, Melville L. 1991. Griphobilharzia amoena n. gen., n. sp.
(Digenea: Schistosomatidae), a parasite of the freshwater crocodile Crocodylus johnstoni (Reptilia: Crocodylia) from Australia, with the erection of a new
subfamily, Griphobilharziinae. Journal of Parasitology 77:65–68.

Popiel I, Basch PF. 1984a. Putative polypeptide transfer from male to female *Schistosoma mansoni*. Molecular and Biochemical Parasitology 11:179–188.

Popiel I, Basch PF. 1984b. Reproductive development of female *Schistosoma mansoni* following bisexual pairing of worms and worm segments. Journal of Experimental Zoology 232:141–150.

Popiel I, Basch PF. 1986. Cholesterol transfer between adult *Schistosoma mansoni*.
Experimental Parasitology 61:343–346.

Popiel I, Cioli D, Erasmus D. 1984a. The morphology and reproductive status of female *Schistosoma mansoni* following separation from male worms. International Journal for Parasitology 14:183–190.

Popiel I, Cioli D, Erasmus D. 1984b. Reversibility of drug-induced regression in female *Schistosoma mansoni* upon pairing with male worms in vivo. Zeitschrift für Parasitenkunde 70:421–424.

Popiel I, Erasmus DA. 1979. *Schistosoma mansoni*: Differential inhibition of vitelline gland morphogenesis induced by selected chemotherapeutic agents. Parasitology 79:xxiii.

Popiel I, Erasmus D. 1981a. *Schistosoma mansoni*: the effect of thiosinamine in vivo and in vitro. Transactions of the Royal Society of Tropical Medicine and Hygiene 75:287–291.

Popiel I, Erasmus DA. 1981b. *Schistosoma mansoni*: Changes in the rate of tyrosine uptake by unisexual females after stimulation by males and male extracts. Journal of Helminthology 55:33–37.

Popiel I, Erasmus DA. 1981c. *Schistosoma mansoni*: Niridazole-induced damage to the vitelline gland. Experimental Parasitology 52:35–48.

Popiel I, Erasmus DA. 1982. *Schistosoma mansoni*: The survival and reproductive status of mature infections in mice treated with oxamniquine. Journal of Helminthology 56:257–261.

Preston AM, El-Khatib SM. 1977. Effect of pregnancy on the number of worms developed in mice infected with *Schistosoma mansoni*. Journal of Parasitology 63:943–945.

Price EW. 1929. A synopsis of the trematode family Schistosomidae, with descriptions of new genera and species. Proceedings of the U.S. National Museum 75 (Article 18):1–39.

Price HF. 1931. Life history of *Schistosomatium douthitti* (Cort). American Journal of Hygiene 13:685–727.

Price Z, Voge M, Beydler S. 1978. Changes in the tegumental surface of *Schistosoma mansoni* during development in the mammalian host. Scanning Electron Microscopy (1978) Vol. II:399–403.

Purnell RE. 1966. Host-parasite relationships in schistosomiasis II.-The effects of age and sex on the infection of mice with cercariae of *Schistosoma mansoni* and of hamsters with cercariae of *Schistosoma haematobium*. Annals of Tropical Medicine and Parasitology 60:94–99.

Race GJ, Martin JH, Moore DV, Larsh JEJ. 1971. Scanning and transmission electronmicroscopy of *Schistosoma mansoni* eggs, cercariae, and adults. American Journal of Tropical Medicine and Hygiene 20:914–924.

Race GJ, Michaels RM, Martin JH, Larsh JE, Matthews JC. 1969. *Schistosoma mansoni* eggs: An electron microscopic study of shell pores and microbarbs. Proceedings of the Society of Experimental Medicine and Biology 130:990–992.

Ramalho-Pinto FJ, Gazzinelli G, Howells RE, Mota-Santos TA, Figueiredo EA, Pellegrino J. 1974. Schistosoma mansoni: Defined system for a step-wise transformation of cercaria to schistosomule in vitro. Experimental Parasitology 36:360–372.

Rao MAN. 1938. A comparative study of *Schistosoma spindalis*, Montgomery, 1906 and *Schistosoma nasalis*, n. sp. Indian Veterinary Journal 14:219–251.

Reis MG, Kuhns J, Blanton R, Davis AH. 1989. Localization and pattern of expression of female specific mRNA in Schistosoma mansoni. Molecular and Biochemical Parasitology 32:113–120.

Reissig M. 1970. Characterization of cell types in the parenchyma of *Schistosoma mansoni*. Parasitology 60, 273–279.

Renaud G, Devidas A, Develoux M, Lamothe F, Bianchi G. 1989. Prevalence of vaginal schistosomiasis caused by Schistosoma haematobium in an endemic village. Transactions of the Royal Society of Tropical Medicine and Hygiene 83:797.

Richards CS. 1975. Genetic factors in susceptibility of Biomphalaria glabrata for different strains of Schistosoma mansoni. Parasitology 70:231–241.

Richards CS. 1976. Variations in infectivity for Biomphalaria glabrata in strains of Schistosoma mansoni from the same geographical area. Bulletin of the World Health Organization 54:706–707.

Richards CS. 1977. Schistosoma mansoni: Susceptibility reversal with age in the snail host Biomphalaria glabrata: Experimental Parasitology 42:165–168.

Richards CS. 1980. Genetic studies on amoebocytic accumulations in Biomphalaria glabrata. Journal of Invertebrate Pathology 35:49–52.

Richards CS. 1984. Influence of snail age on genetic variations in susceptibility of Biomphalaria glabrata for infection with Schistosoma mansoni. Malacologia 25:493–502.

Richards CS, Shade PL. 1987. The genetic variation of compatibility in Biomphalaria glabrata and Schistosoma mansoni. Journal of Parasitology 73:1146–1151.

Rieger RM. 1985. The phylogenetic status of the acoelomate organization within the Bilateria: a histological perspective. Pp. 101–122 In: SC Morris, JD George, R Gibson, and HM Platt, Eds. The Origins and Relationships of Lower Invertebrates. Oxford. Clarendon Press.

Robinson EJ. 1957a. A possible effect of testosterone on the development of *Schistosoma mansoni*. Journal of Parasitology 43:59.

Robinson EJ Jr. 1957b. Further studies on the effect of abnormal host metabolism on *Schistosoma mansoni*. Journal of Parasitology 45:295–299.

Robinson EJ Jr. 1959. Recovery of *Schistosoma mansoni* from hormonally imbalanced hosts. Experimental Parasitology 8:236–243.

Rogers SH, Bueding E. 1975. Anatomical localization of glucose uptake by *Schistosoma mansoni* adults. International Journal for Parasitology 5:369–371.

Rohde K. 1977. The bird *Schistosome Austrobilharzia* terrigalensis from the Great Barrier Reef, Australia. Zeitschrift für Parasitenkunde 52:39–51.

Rollinson D. 1985. Biochemical genetics in the study of schistosomes and their intermediate hosts. Parassitologia 27:123–139.

Rollinson D, Imbert-Establet D, Ron GC. 1986. Schistosoma mansoni from naturally infected Rattus rattus in Guadeloupe: identification, prevalence and enzyme polymorphism. Parasitology 93:39–53.

Rollinson D, Simpson AJG, Eds. 1987. The Biology of Schistosomes From Genes to Latrines. London. Academic Press.

Rollinson D, Southgate VR. 1987. The genus Schistosoma: A taxonomic appraisal. Pp. 1–49 In: D Rollinson, AJG Simpson, Eds. The Biology of Schistosomes From Genes to Latrines. London. Academic Press.

Rollinson D, Southgate VR, Vercruyse J, Moore PJ. 1990. Observations on natural and experimental interactions between Schistosoma bovis and S. curassoni from West Africa. Acta Tropica 47:101–114.

Rombert PC, Ordens ML. 1978. Influencia do sexo do hospedeiro na susceptibildade a infecção por *Schistosoma mansoni*. Anais do Instituto de Higiene e Medicina Tropical 5:363–366.

Ross GC, Southgate VR, Knowles RJ. 1978. Observations on some isoenzymes of strains of *Schistosoma bovis, S. mattheei, S. margrebowiei,* and *S. lieperi.* Zeitschrift für Parasitenkunde 57:49–56.

Rotmans JP, Burgers A. 1984. Analysis of sex-linked *Schistosoma mansoni* antigens by immunoprecipitation and immunoblot techniques. Tropical and Geographical Medicine 36:383.

Rowntree S, James C. 1977. Single sex cercariae of *Schistosoma mansoni*: A comparison of male and female infectivity. Journal of Helminthology 51:69–70.

Ruff MD, Davis GM, Werner JK. 1973. Schistosoma japonicum: Disk electrophoretic protein patterns of the Japanese, Philippine, and Formosan strains. Experimental Parasitology 33:437–446.

Ruiz JM, Coelho E. 1952. Schistosomose experimental. 2. Hermafroditismo do *Schistosoma mansoni* verificado na cobaiá. Memorias do Instituto Butantan 24:115–126.

Rumjanek FD. 1989. Regulation of gene expression in the Schistosoma mansoni female. Memorias do Instituto Oswaldo Cruz 84 Suppl 1:197–198.

Rumjanek FD, Braga VM, Silveira AM, Rabelo EM, Campos EG, Rodriguez V. 1987. Genomic regulation in immature females of Schistosoma mansoni. Memorias do Instituto Oswaldo Cruz 84 Suppl 4:209–211.

Rumjanek FD, Simpson AJ. 1980. The incorporation and utilization of radiolabeled lipids by adult *Schistosoma mansoni* in vitro. Molecular and Biochemical Pharmacology 1:31–44.

Ruppel A, Cioli D. 1977. A comparative analysis of various developmental stages of *Schistosoma mansoni* with respect to their protein composition. Parasitology 75:339–343.

Sagawa E, Ogi K, Sumikoshi Y. 1928. Study concerning the influence of sex on the growth of animal bodies. First report. Experiment on *Schistosoma japonicum*. Transactions of the Japanese Pathology Society 18:494–500.

Sahba GH, Malek EA. 1977a. Hermaphroditic female Heterobilharzia americana. Journal of Parasitology 63:947–948.

Sahba GH, Malek EA. 1977b. Unisexual infections with *Schistosoma haematobium* in the mouse. American Journal of Tropical Medicine and Hygiene 26:331–333.

Sakamoto K, Ishii Y. 1976. Fine structure of schistosome eggs as seen through the scanning electron microscope. American Journal of Tropical Medicine and Hygiene 25:841–844.

Sakamoto K, Ishii Y. 1977. Scanning electron microscope observations on adult *Schistosoma japonicum*. Journal of Parasitology 63:407–412.

Saladin KS. 1979. Behavioral parasitology and perspectives on miracidial host-finding. Zeitschrift für Parasitenkunde 60:197–210.

Salafsky B, Fusco AC, Whitley K, Nowicki D, Ellenberger B. 1988. Schistosoma mansoni: analysis of cercarial transformation methods. Experimental Parasitology 67:116–127.

Samuelson JC, Caulfield JP. 1985. The cercarial glycocalyx of Schistosoma mansoni. Journal of Cell Biology 100:1423–1434.

Samuelson JC, Quinn JJ, Caulfield JP. 1984. Hatching, chemokinesis and transformation of miracidia of Schistosoma mansoni. Journal of Parasitology 70:321–331.

Samuelson JD, Stein LD. 1989. Schistosoma mansoni: Increasing saline concentration signals cercariae to transform to schistosomula. Experimental Parasitology 69:23–29.

Saoud MFA. 1964. A preliminary note on hermaphroditic males of *Schistosoma bovis*. Transactions of the Royal Society of Tropical Medicine and Hygiene 58, 288–289.

Saoud MFA. 1965. Comparative studies on the characteristics of some geographical strains of *Schistosoma mansoni* in mice and hamsters. Journal of Helminthology 39:101–112.

Saoud MFA. 1966a. On the morphology of *Schistosoma rodhaini* from Kenya. Journal of Helminthology 40:147–154.

Saoud MFA. 1966b. The infectivity and pathogenicity of geographical strains of *Schistosoma mansoni*. Transactions of the Royal Society of Tropical Medicine and Hygiene 60, 585–600.

Saoud MFA. 1966c. On the infraspecific variations of the male sexual glands of *Schistosoma mansoni*. Journal of Helminthology 40:385–394.

Sathe BD, Mukherji S, Gaitonde BB, Renapurkar DM. 1981. Reinvestigation of an old focus of schistosomiasis in Gimvi Village, District Ratnagiri, in Maharashtra State, India. Bulletin of the Haffkine Institute 9:34–37.

Schiller EL, Bueding E, Turner VM, Fisher J. 1975. Aerobic and anaerobic carbohydrate metabolism and egg production of *Schistosoma mansoni* in vitro. Journal of Parasitology 61:385–389.

Schnitzer B, Sodeman T, Sodeman WAJ, Durkee T. 1971. Microspines on *Schistosoma japonicum* and *S. haematobium* egg shells. Parasitology 62:385–387.

Schutte HJ. 1974a. Studies on the South African strain of Schistosoma mansoni. Part 1: Morphology of the miracidium. South African Journal of Science 70:299–302.

Schutte HJ. 1974b. Studies on the South African strain of Schistosoma mansoni. Part 2: The intra-molluscan larval stages. South African Journal of Science 70:327–346.

Schutte HJ. 1975. Studies on the South African strain of Schistosoma mansoni. Part 3: Notes on certain host-parasite relations between intra-molluscan larvae and intermediate host. South African Journal of Science 71:8–20.

Schwanbek A, Becker W, Rupprecht H. 1986. Quantification of parasite development in the host-parasite system Biomphalaria glabrata and schistosoma mansoni. Zeitschrift für Parasitenkunde 72:365–373.

Schwetz J. 1951. A comparative morphological and biological study of *Schistosoma haematobium, S. bovis, S. intercalatum* Fisher, 1934, *S. mansoni* and *S. rodhaini* Brunpt, 1931. Annals of Tropical Medicine and Parasitology 45:92–98.

Schwetz J. 1953. Sur les variations individuelles dans les oeufs des Schistosomes. Comptes Rendus des Séances de la la Société de Biologie et de ses Filiales (Paris), 147:2051–2053.

Schwetz J. 1954. On two schistosomes from wild rodents of the Belgian Congo: *Schistosoma rodhaini* Brumpt, 1931; and Schistosoma mansoni var. rodentorum Schwetz, 1953; and their relationship to *S. mansoni* of man. Transactions of the Royal Society of Tropical Medicine and Hygiene 48:89–100.

Seed JL, Boff M, Bennett JL. 1978. Phenol oxidase activity: Induction in female schistosomes by in vitro incubation. Journal of Parasitology 64:283–289.

Senft AW. 1968. Studies in proline metabolism by *Schistosoma mansoni*. I. Radioautography following in vitro exposure to radio-proline C14. Comparative Biochemistry and Physiology 27:251–261.

Senft AW, Gibler WB, Knopf PM. 1978. Scanning electron microscope observations on tegument maturation in *Schistosoma mansoni* grown in permissive and non-permissive hosts. American Journal of Tropical Medicine and Hygiene 27:258–266.

Severinghaus AE. 1928. Sex studies on *Schistosoma japonicum*. Quarterly Journal of Microscopical Science 71:653–702.

Shaw JR. 1977. *Schistosoma mansoni:* Pairing in vitro and development of females from single sex infections. Experimental Parasitology 41:54–65.

Shaw JR, Erasmus DA. 1981. *Schistosoma mansoni:* An examination of the reproductive status of females from single sex infections. Parasitology 82:121–124.

Shaw JR, Marshall I, Erasmus DA. 1977. *Schistosoma mansoni:* In vitro stimulation of vitelline cell development by extracts of male worms. Experimental Parasitology 42:14–20.

Shaw, MK. 1987. Schistosoma mansoni: vitelline gland development in females from single sex infections. Journal of Helminthology 61:253–259.

Shaw MK, Erasmus DA. 1982. *Schistosoma mansoni:* The presence and ultrastructure of vitelline cells in adult males. Journal of Helminthology 56:51–53.

Sher A, James S, Correa-Oliveira R, Hieny S, Pearce E. 1989. Schistosome vaccines: current progress and future prospects. Parasitology 98 (Suppl):S61–S68.

Shirazian D, Schiller EL. 1982. Mating recognition by *Schistosoma mansoni* in vitro. Journal of Parasitology 68:650–652.

Shoop WL. 1988. Trematode transmission patterns. Journal of Parasitology 74:46–59.

Short RB. 1948a. Hermaphrodites in a Puerto Rican strain of *Schistosoma mansoni*. Journal of Parasitology 34:240–242.

Short RB. 1948b. Inter-generic crosses among schistosomes (Trematoda: Schistosomatidae). Journal of Parasitology 34 (6 part 2):30.

Short RB. 1951. Hermaphroditic female *Schistosomatium douthitti* (Trematoda: Schistosomatidae). Journal of Parasitology 37:547–555.

Short RB. 1952a. Sex studies on *Schistosomatium douthitti* (Cort 1914) Price 1931 (Trematoda: Schistosomatidae). American Midland Naturalist 47:1–54.

Short RB. 1952b. Uniparental miracidia of *Schistosomatium douthitti* and their progeny (Trematoda: Schistosomatidae). American Midland Naturalist 48:55–68.

Short RB. 1957. Chromosomes and sex in *Schistosomatium douthitti*. Journal of Heredity 48:2–6.

Short RB. 1983. Sex and the single schistosome. Journal of Parasitology 69:4–22.

Short RB, Liberatos JB, Teehan WH, Bruce JI. 1989. Conventional Giemsa-stained and C-banded chromosomes of seven strains of Schistosoma mansoni. Journal of Parasitology 75:920–927.

Short RB, Menzel MY. 1959. Chromosomes in parthenogenetic miracidia and embryonic cercariae of *Schistosomatium douthitti*. Experimental Parasitology 8:249–264.

Short RB, Menzel MY, Pathak S. 1979. Somatic chromosomes of *Schistosoma mansoni*. Journal of Parasitology 65:471–473.

Short RB, Teehan WH, Liberatos JD. 1987. Chromosomes of Heterobilharzia americana (Digenea: Schistosomatidae), with ZWA sex determination, from Louisiana. Journal of Parasitology 73:941–946.

Siefker C, Pax RA, Bennett JL. 1983. *Schistosoma mansoni:* A comparison of male and female muscle physiology. Comparative Biochemistry and Physiology 76C:377–382.

Siegel DA, Tracy JW. 1988. Effect of pairing in vitro on the glutathione level of male Schistosoma mansoni. Journal of Parasitology 74:524–531.

Siegel DA, Tracy JW. 1989. Schistosoma mansoni: influence of the female parasite on glutathione biosynthesis in the male. Experimental Parasitology 69:116–124.

Silk MH, Spence IM. 1969a. Ultrastructural studies of the blood fluke *Schistosoma mansoni* II. The musculature. South African Journal of Medical Sciences 34:11–20.

Silk MH, Spence IM. 1969b. Ultrastructural studies of the blood fluke *Schistosoma mansoni* III. The nerve and sensory structures. South African Journal of Medical Science 34:93–104.

Silk MH, Spence IM, Buch B. 1970. Observations of S. mansoni blood flukes in the scanning electron microscope. South African Journal of Medical Sciences 35:93–112.

Silk MH, Spence IM, Gear JHS. 1969. Ultrastructural studies of the blood fluke *Schistosoma mansoni* I. The integument. South African Journal of Medical Sciences 34:1–10.

Silveira AM, Friche AA, Rumjanek FD. 1986. Transfer of [14C] cholesterol and its metabolites between adult male and female worms of Schistosoma mansoni. Comparative Biochemistry and Physiology 85B:851–857.

Simpson AJG. 1987. Schistosome molecular biology. Pp. 147–161 In: D Rollinson, AJG Simpson, Eds. The Biology of Schistosomes From Genes to Latrines. London. Academic Press.

Simpson AJ, Chaudri M, Knight M, Kelly C, Rumjanek F, Martin S, Smithers SR. 1987. Characterisation of the structure and expression of the gene encoding a major female specific polypeptide of Schistosoma mansoni. Molecular and Biochemical Parasitology 22:169–176.

Simpson AJG, Knight M. 1986. Cloning of a major developmentally regulated gene expressed in mature females of *Schistosoma mansoni*. Molecular and Biochemical Parasitology 18:25–35.

Simpson AJG, Sher A, McCutchan TF. 1982. The genome of *Schistosoma mansoni*: isolation of DNA, its size, bases and repetitive sequences. Molecular and Biochemical Parasitology 6:125–137.

Sinha PK, Srivastava HD. 1956. Studies on Schistosoma incognitum Chandler, 1926 I. On the synonymy and morphology of the blood-fluke. Parasitology 46:91–100.

Sinha PK, Srivastava HD. 1960. Studies on Schistosoma incognitum Chandler, 1926. II. On the life history of the blood-fluke. Journal of Parasitology 46:629–641.

Skriabin KI. 1951. Trematody Zhivotnykh i Cheloveka. Moskva. Istadylest'vo Akad Nauk SSSR. Tom V.

Sminia T. 1981. Gastropods. Pp 191–232 In: NA Ratcliffe, AF Rowley Eds. Invertebrate Blood Cells. Volume 1. General Aspects, Animals Without True Circulatory Systems to Cephalopods. London. Academic Press.

Smith, JW. 1972. The blood flukes (Digenea: Sanguinicolidae and Spirorchidae) of cold-blooded vertebrates and some comparisons with the schistosomes. Helminthological Abstracts 41A:161–204.

Smith M. 1974. Radioassays for the proteolytic enzymes secreted by living eggs of Schistosoma mansoni. International Journal for Parasitology 4:681–683.

Smith M, Clegg JA, Webbe G. 1976. Culture of *Schistosoma haematobium* in vivo and in vitro. Annals of Tropical Medicine and Parasitology 70, 101–107.

Smith MA, Clegg JA. 1979. Different levels of immunity to *Schistosoma mansoni* in the mouse: the role of variant cercariae. Parasitology 78:311–321.

Smith SWG, Chappell LH. 1990. Single sex schistosomes and chemical messengers (letter). Parasitology Today 6:297–298.

Smith TM, Brooks TJJ, Lockard VG. 1970. In vitro studies on cholesterol metabolism in the blood fluke *Schistosoma mansoni*. Lipids 5:854–856.

Smith TM, Doughty BL, Brown JN. 1977. Fatty acid and lipid class composition of *Schistosoma japonicum* Comparative Biochemistry and Physiology 57B:59–63.

Smith TM, Lucia HL, Doughty BL, von Lichtenberg FC. 1971. The role of phospholipids in schistosome granulomas. Journal of Infectious Diseases 123:629–639.

Smithers SR, Simpson AJG, Yi X, Omer-Ali P, Kelly C, McLaren DJ. 1987. The mouse model of schistosome immunity. Acta Tropica 44 (Suppl 12):21–30.

Smithers SR, Terry RJ. 1965. The infection of laboratory hosts with cercariae of Schistosoma mansoni and the recovery of the adult worms. Parasitology 55:695–700.

Smyth JD, Halton DW. 1983. The Physiology of Trematodes. Second Edition. Cambridge. Cambridge Univ. Press.

Sobhon P, Upatham ES, Koonchornboon T, Saitongdee P, Khunborivan V, Yuan HC, Vongpayabal P, Ow-Yang CK, Greer G. 1983. Microtropography of the surface of adult *Schistosoma japonicum*-like (Malaysian) as observed by scanning electron microscopy. Southeast Asian Journal of Tropical Medicine and Public Health 14:439–450.

Sogandares-Bernal F. 1966. Studies in American paragonimiasis. IV. Observations on the pairing of adult worms in laboratory infections of domestic cats. Journal of Parasitology 52, 701–703.

Southgate VR, Knowles RJ. 1977. On *Schistosoma margrebowiei* Le Roux, 1933: The morphology of the egg, miracidium and cercaria, the compatibility with the species of Bulinus, and the development in Mesocricetus auratus. Zeitschrift für Parasitenkunde 54:233–250.

Southgate VR, Rollinson D. 1987. Natural history of transmission and schistosome infection. Pp. 347–378 In: D. Rollinson, AJG Simpson, Eds. The Biology of Schistosomes From Genes to Latrines. London. Academic Press.

Southgate VR, Rollinson D, Ross GC, Knowles RJ. 1982. Mating behaviour in mixed infections of *Schistosoma maematobium* and *S. intercalatum*. Journal of Natural History 16:491–496.

Southgate VR, Rollinson D, Ross GC, Knowles RJ, Vercruysse J. 1985. On Schistosoma curassoni, S. haematobium and S. bovis from Senegal, development in Mesocricetus auratus, compatibility with species of Bulinus, and their enzymes. Journal of Natural History 19:1249–1267.

Southgate VR, Ross GC, Knowles RJ. 1981. On *Schistosoma leiperi* Le Roux 1955: Scanning electron microscopy of adult worms, compatibility with species of *Bulinus,* development in *Mesocricetus auratus,* and isoenzymes. Zeitschrift für Parasienkunde 66:63–81.

Southgate VR, van Wijk HB, Wright CA. 1976. Schistosomiasis at Loum, Cameroun; *Schistosoma maematobium, S. intercalatum* and their natural hybrid. Zeitschrift für Parasitenkunde 49:145–159.

Spence IM, Silk MH. 1970. Ultrastructural studies of the blood fluke—*Schistosoma mansoni*. IV. The digestive system. South African Journal of Medical Sciences 35:93–112.

Spence IM, Silk MH. 1971a. Ultrastructural studies of the blood fluke—*Schistosoma mansoni*. V. The female reproductive system—a preliminary report. South African Journal of Medical Sciences 36:41–50.

Spence IM, Silk MH. 1971b. Ultrastructural studies of the blood fluke—*Schistosoma mansoni*. VI. The Mehlis gland. South African Journal of Medical Sciences 36:69–76.

Spotila LD, Rekosh DM, Boucher JM, LoVerde PT. 1987. A cloned DNA probe identifies the sex of Schistosoma mansoni cercariae. Molecular and Biochemical Parasitology 26:17–20.

Standen OD. 1949. Experimental schistosomiasis. II. Maintenance of *Schistosoma mansoni* in the laboratory with some notes on experimental infection with *S. haematobium*. Annals of Tropical Medicine and Parasitology 43:268–283.

Standen OD. 1953a. The relationship of sex in *Schistosoma mansoni* to migration within the hepatic portal system of experimentally infected mice. Annals of Tropical Medicine and Parasitology 47:139–145.

Standen OD. 1953b. Experimental schistosomiasis. III. Chemotherapy and mode of drug action. Annals of Tropical Medicine and Parasitology 47:26–43.

Standen OD. 1955. The treatment of experimental schistosomiasis in mice: sexual maturity and drug response. Annals of Tropical Medicine and Parasitology 49:183–192.

Stenger RJ, Warren KS, Johnson EA. 1967. An ultrastructural study of hepatic granulomas and schistosome egg shells in murine hepatosplenic schistosomiasis. Experimental and Molecular Pathology 7:116–132.

Stirewalt M. 1951. The frequency of bisexual infections of *Schistosoma mansoni* in snails of the species Australorbis glabratus (Say). Journal of Parasitology 37:42–47.

Stirewalt MA. 1974. Schistosoma mansoni: Cercariae to schistosomule. Advances in Parasitology 12:115–182.

Stirewalt MA, Fregeau WA. 1965. Effect of selected experimental conditions on penetration and maturation of *Schistosoma mansoni* in mice. I. Environmental. Experimental Parasitology 17:168–179.

Stirewalt MA, Fregeau WA. 1968. Effect of selected experimental conditions on penetration and maturation of cercariae of *Schistosoma mansoni* in mice. II. Parasite-release conditions. Experimental Parasitology 22:73–95.

Stirewalt MA, Kuntz RE, Evans AS. 1951. The relative susceptibilities of the commonly-used laboratory mammals in infection by *Schistosoma mansoni*. American Journal of Tropical Medicine and Hygiene 31:57–82.

Stjernholm RL, Warren KS. 1974. *Schistosoma mansoni*: Utilization of exogenous metabolites by eggs in vitro. Experimental Parasitology 36:222–232.

Striebel HP. 1969. The effects of niridazole in experimental schistosomiasis. Annals of the New York Academy of Sciences 160:491–518.

Striebel HP, Kradolfer F. 1966. Mode of action of CIBA 32655-Ba in experimental schistosomiasis. Acta Tropica Suppl 9:54–57.

Stunkard HW. 1970. Trematode parasites of insular and relict vertebrates. Journal of Parasitology 56:1041–1054.

Stunkard HW. 1975. Life-histories and systematics of parasitic flatworms. Systematic Zoology 24:378–385.

Stunkard HW, Hinchcliffe MC. 1952. The morphology and life-history of Microbilharzia variglandis (Miller and Northrup, 1926) Stunkard and Hinchcliffe 1951, avian blood flukes whose larvae cause "swimmer's itch" of ocean beaches. Journal of Parasitology 38:248–265.

Sturrock RF. 1966. Daily egg output of schistosomes. Transactions of the Royal Society of Tropical Medicine and Hygiene 60:139–140.

Sturrock RF. 1986. A review of the use of primates in studying human schistosomiasis, Journal of Medical Primatology 15:267–279.

Sturrock RF, Cottrell BJ, Lucas S, Reid W, Seitz HM, Wilson RA. 1988. Observations on the implications of pathology induced by experimental schistosomiasis in baboons in evaluating the development of resistance to challenge infection. Parasitology 96:37–48.

Sugiura S. 1934. Studies on the sex of *Schistosoma japonicum* I, II. Japanese Journal of Zoology 5(Abstracts):23.

Sullivan JT, Richards CS. 1982. Increased susceptibility to infection with Ribeiroia marini in Biomphalaria glabrata exposed to X-rays. Journal of Invertebrate Pathology 40:303–304.

Suyemori S. 1922. Development in the final host of the Formosan *Schistosoma japonicum*. Taiwan Igakkai Zasshi 220:1–24.

Tada S. 1928. Ueber die Entwicklung und den Bau des *Schistosomum japonicum* in Endwirte. Okayama Igakkai Zasshi 40:1827–1868.

Tanabe B. 1923. The life history of a new schistosome, Schistosomatium pathlocopticum Tanabe, found in experimentally infected mice. Journal of Parasitology 9:183–198.

Tanabe K. 1919. A contribution to the morphology and development of *Schistosoma japonicum*. Igaku Chuo Zasshi 17:6.

Tanaka M, Matsuda H. 1973. Statistical analysis on the probability of uni- and bisexual infections of *Schistosoma japonicum* in *Oncomelania nosophora*. Japanese Journal of Experimental Medicine 43:423–434.

Taylor MG. 1970. Hybridisation experiments on five species of African schistosomes. Journal of Helminthology 44:253–314.

Taylor MG. 1971. Further observations on the sexual maturation of schistosomes in single-sex infections. Journal of Helminthology 45:89–92.

Taylor MG, Amin MBA, Nelson GS. 1969. "Parthenogenesis" in *Schistosoma mattheei*. Journal of Helminthology 43:197–206.

Tewari HC, Singh KS. 1979. Pathogenesis of *Schistosoma incognitum* in mice with special reference to the mechanism of anaemia. Indian Journal of Animal Science 49:380–383.

Théron A. 1989. Hybrids between Schistosoma mansoni and Schistosoma rodhaini. Characterization by cercarial emergence rhythms. Parasitology 99:225–228.

Théron A, Combes C. 1983. Analyse génétique du rythme d'émergence des cercaires de *Schistosoma mansoni* par croissement de souches à pics d'émission précoces ou tardifs. Comptes Rendus de l'Academie des Sciences III 297:571–574.

Théron A, Jourdane J. 1979. Séquence de reconversion des sporocystes de Schistosoma mansoni producteurs de cercaires en vue de la production de nouvelles générations de sporocystes. Zeitschrift für Parasitenkunde 61:63–71.

Thompson DP, Pax RA, Benett JL. 1984. Schistosoma mansoni: A comparative study of schistosomula transferred mechanically and by skin penetration. Electrophysiological responses to a wide range of substances. Parasitology 88:477–489.

Thompson SN. 1987. Effect of Schistosoma mansoni on the gross lipid composition of its vector Biomphalaria glabrata. Comparative Biochemistry and Physiology 87:357–360.

Timms AR, Bueding E. 1959. Studies of a proteolytic enzyme from *Schistosoma mansoni*. British Journal of Pharmaceutics and Chemotherapy 14:68–73.

Tomosky TK, Bennett JL, Bueding E. 1974. Tryptaminergic and dopaminergic responses of *Schistosoma mansoni*. Journal of Pharmacology and Experimental Therapeutics 190:260–271.

Toro-Goyco E, Rosas del Valle M. 1970. *Schistosoma mansoni*. 1. Chemical composition of eggs. Experimental Parasitology 27:265–272.

Torpier G, Hirn M, Nirde P, DeReggi M, Capron A. 1982. Detection of ecdysteroids in the human trematode Schistosoma mansoni. Parasitology 84:123–130.

Tulloch GS, Kuntz RE, Davidson DL, Huang TC. 1977. Scanning electron microscopy of the integument of *Schistosoma mattheei* Veglia and Le Roux, 1929. Transactions of the American Microscopical Society 96:41–47.

Uglem GL, Read CP. 1975. Sugar transport and metabolism in *Schistosoma mansoni*. Journal of Parasitology 61:390–397.

Ulmer MJ, vandeVusse FJ. 1970. Morphology of *Dendritobilharzia pulverulenta* (Braun, 1901) Skrjabin, 1924 (Trematoda: Schistosomatidae) with notes on secondary hermaphroditism. Journal of Parasitology 56:67–74.

Upatham ES. 1972a. Rapidity and duration of hatching of St. Lucian *Schistosoma mansoni* eggs in outdoor habitats. Journal of Helminthology 46:271–276.

Upatham ES. 1972b. Studies on the hatching of *Schistosoma mansoni* eggs in the standing-water and running-water habitats in St. Lucia, West Indies. Southeast Asian Journal of Tropical Medicine and Public Health 3:600–604.

Upatham ES. 1973a. The effects of water temperature on the penetration and development of St. Lucian *Schistosoma mansoni* in local *Biomphalaria glabrata*. Southeast Asian Journal of Tropical Medicine and Public Health 4:367–370.

Upatham ES. 1973b. Location of Biomphalaria glabrata (Say) by miracidia of Schistosoma mansoni Sambon in natural standing and running waters on the West Indian island of St. Lucia. International Journal for Parasitology 3:289–297.

Upatham ES. 1974. Infectivity of Schistosoma mansoni cercariae in natural St. Lucian habitats. Annals of Tropical Medicine and Parasitology 68:235–236.

Valadares TE, Coelho PM, Pellegrino J, Sampaio IB. 1981a. *Schistosoma mansoni*: Comparação da oviposição entre as cepas LE (Belo Horizonte), SP (São Paulo) e ST (Liberia) em camundongos. Revista do Instituto de Medicina Tropical de São Paulo 23:1–5.

Valadares TE, Coelho PM, Pellegrino J, Sampaio IB. 1981b. *Schistosoma mansoni*: Aspectos da oviposição da cepa LE em camundongos infectados com um casal de vermes. Revista do Instituto de Medicina Tropical de São Paulo 31: 6–11.

Valadares TE, Coelho PM, Sampaio IB. 1980. *Schistosoma mansoni*: Aspectos da oviposição (distribuição de ovos nos intestinos e fígado de camundungos e eliminação destes ovos pelos fezes) das cepas LE e CA. Revista Brasileira de Malariologia e Doenças Tropicais 32:53–59.

van Rensburg LJ, van Wyk JA. 1981. Studies on schistosomiasis. 10. Development of *Schistosoma mattheei* in sheep infested with equal numbers of male and female cercariae. Onderstepoort Journal of Veterinary Research 48:77–86.

Vermund SH, Bradley DJ, Ruiz-Tiben E. 1983. Survival of *Schistosoma mansoni* in the human host: Estimates from a community-based prospective study in Puerto Rico. American Journal of Tropical Medicine and Hygiene 32:1040–1048.

van der Knaap WPW, Loker ES. 1990. Immune mechanisms in trematode-snail interactions. Parasitology Today 6:175–182.

Vignali DA, Bickle QD, Taylor MG. 1989. Immunity to Schistosoma mansoni in vivo: Contradiction or clarification? Immunology Today 10:410–416.

Voge M, Bruckner D, Bruce JI. 1978a. *Schistosoma mekongi,* sp.n. from man and animals, compared with four geographic strains of *Schistosoma japonicum.* Journal of Parasitology 64:577–584.

Voge M, Price Z, Jansma WB. 1978b. Observations on the surface of different strains of adult *Schistosoma japonicum*. Journal of Parasitology 64:368–372.

Voge M, Price Z, Bruckner DA. 1978c. Changes in tegumental surface of *Schistosoma mekongi* Voge, Bruckner, and Bruce 1978, in the mammalian host. Journal of Parasitology 64:944–947.

Vogel H. 1941a. Infektionsversuche an verschiedenen Bilharzia-Zwischenwirten mit einem einzelnen Mirazidium von *Bilharzia mansoni* und *B. japonica*. Zentralblatt für Bakteriologie Abteilung I. Originale 148:29–35.

Vogel H. 1941b. Ueber den Einfluss des Geschlechtspartners auf Wachstum und Entwicklung bei *Bilharzia mansoni* und Kreuzpaarungen zwischen verscheidenen Bilharzia-Arten. Zentralblatt für Bakteriologie Abteilung I. Originale 148:78–96.

Vogel H. 1942a. Über Entwicklung, Lebensdauer und Tod der Eier von *Bilharzia japonica* in Wirtsgewebe. Deutsche Tropenmedizin Zeitschrift 46:57–69; 81–91.

Vogel H. 1942b. Ueber die Nachkommenschaft aus Kreuzpaarungen zwischen *Bilharzia mansoni* und *B. japonica*. Zentralblatt für Bakteriologie Abteilung I. Originale 149:319–333.

Vogel H. 1947. Hermaphrodites of *Schistosoma mansoni*. Annals of Tropical Medicine and Parasitology 4:266–277.

Vogel H, Minning W. 1947. Ueber die Einwirkung von Brechweinstein, Fuadin und Emetin auf *Bilharzia japonica* und deren Eier im Kaninchenversuch. Acta Tropica 4:21–56; 97–116.

Walker TK, Rollinson D, Simpson AJG. 1986. Differentiation of *Schistosoma haematobium* from related species using cloned ribosomal RNA gene probes. Molecular and Biochemical Parasitology 20:123–131.

Walker TK, Rollinson D, Simpson AJ. 1989a. A DNA probe from Schistosoma mansoni allows rapid determination of the sex of larval parasites. Molecular and Biochemical Parasitology 33:93–100.

Walker TK, Simpson AJ, Rollinson D. 1989b. Differentiation of Schistosoma mansoni from Schistosoma rodhaini using cloned DNA probes. Parasitology 98:75–80.

Wang CG, Zhu QY, Hang PY, Zhu YW, Wang JW, Shen YP, Yang SJ, Chao JY. 1984. HLA and schistosomiasis japonica. Chinese Medical Journal 97:603–605.

Wang FL, Su YF, Yang GM, Wang XZ, Qui ZY, Zhou XK, Hu ZQ. 1986. Isoenzymes of phenol oxidase in adult female *Schistosoma japonicum*. Molecular and Biochemical Parasitology 18:69–72.

Ward RD, Lewis FA, Yoshino TP, Dunn TS. 1988. Schistosoma mansoni: relationship between cercarial production levels and snail host susceptibility. Experimental Parasitology 66:78–85.

Warren KS. 1967. A comparison of Puerto Rican, Brazilian, Egyptian and Tanzanian strains of *Schistosoma mansoni* in mice: Penetration of cercariae, maturation of schistosomes and production of liver disease. Transactions of the Royal Society of Tropical Medicine and Hygiene 61:795–802.

Warren KS. 1973. Schistosomiasis: The Evolution of a Medical Literature. Selected Abstracts and Citations, 1852–1972. Cambridge, Massachusetts. MIT Press.

Warren KS. 1980. Schistosome eggs as antigenic depots: Production, secretion, destruction. In: L. Israel, P.H. Lagrange, and J.C. Solomon, Eds. Immunologie Antitumorale et Antiparasitaire 449–460.

Warren KS, Berry EG. 1972. Induction of hepatosplenic disease by single pairs of the Philippine, Formosan, Japanese, and Chinese strains of *Schistosoma japonicum*. Journal of Infectious Diseases 126:482–491.

Warren KS, Hoffman DB Jr. 1976. Schistosomiasis III. Abstracts of the Complete Literature, 1963–1974. 2 Volumes. Washington, DC. Hemisphere Publishing Co.

Warren KS, Mahmoud AAF, Cummings P, Murphy DJ, Houser HB. 1974. *Schistosoma mansoni* in Yemeni in California: Duration of infection, presence of disease, therapeutic management. American Journal of Tropical Medicine and Hygiene 23:902–909.

Warren KS, Newill VA. 1967. Schistosomiasis: A Bibliography of the World's Literature from 1852 to 1962. Cleveland, Ohio. 2 Volumes. Press of Western Reserve University.

Warren KS, Peters PA. 1967. Comparison and maturation of *Schistosoma mansoni* in the hamster, mouse, guinea pig, rabbit, and rat. American Journal of Tropical Medicine and Hygiene 16:718–722.

Watt G, Sy N. 1988. HLA-Typing in Schistosoma japonicum infection (letter). Transactions of the Royal Society of Tropical Medicine and Hygiene 82:350.

Webbe G, James C. 1971a. Intraspecific variation of *Schistosoma haematobium*. Journal of Helminthology 45:403–413.

Webbe G, James C. 1971b. A comparison of two geographical strains of *Schistosoma haematobium*. Journal of Helminthology 45:271–284.

Webster P, Mansour TE, Bieber D. 1989. Isolation of a female-specific highly repeated Schistosoma mansoni DNA probe and its use in an assay of cercarial sex. Molecular and Biochemical Parasitology 36:217–222.

Wharton DA. 1983. The production and functional morphology of helminth egg shells. Parasitology. 86 Suppl:85–97.

Wheater PR, Wilson RA. 1979. *Schistosoma mansoni*: A histological study of migration in the laboratory mouse. Parasitology 79:49–62.

Whitfield PJ, Evans NA. 1983. Parthenogenesis and asexual multiplication among parasitic helminths. Parasitology 86:121–160.

Wilson RA. 1987. Development and migration in the mammalian host. Pp. 115–146 In: D Rollinson, AJG Simpson, Eds. The Biology of Schistosomes From Genes to Latrines. London. Academic Press.

Wilson RA, Coulson PS. 1986. *Schistosoma mansoni*: dynamics of migration through the vascular system of the mouse. Parasitology 92:83–100.

Wilson RA, Draskau T, Miller P, Lawson JR. 1978. *Schistosoma mansoni*: The activity and development of the schistosomulum during migration from the skin to the hepatic portal system. Parasitology 77:57–73.

Wishahi M, el-Baz HG, Shaker ZA. 1989. Association between HLA-A, B,C and DR antigens and clinical manifestations of Schistosoma haematobium in the bladder. European Urology 16:138–143.

Witenberg G, Lengy J. 1967. Redescription of Ornithobilharzia canaliculata (Rud.) Odhner, with notes on classification of the genus Ornithobilharzia and the subfamily Schistosomatinae (Trematoda). Israel Journal of Zoology 16:193–204.

World Health Organization. 1960. Bibliography on Bilharziasis 1949–1958. Geneva World Health Organization.

Wright CA. 1971. Flukes and Snails. London. Allen and Unwin.

Wright CA, Bennett MS. 1967a. Studies on *Schistosoma haematobium* in the laboratory. 1. A strain from Durban, Natal, South Africa. Transactions of the Royal Society of Tropical Medicine and Hygiene 61:221–227.

Wright CA, Bennett MS. 1967b. Studies on *Schistosoma haematobium* in the laboratory. 2. A strain from South Arabia. Transactions of the Royal Society of Tropical Medicine and Hygiene 61:228–233.

Wright CA, Knowles RJ. 1972. Studies on *Schistosoma haematobium* in the laboratory. III. Strains from Iran, Mauritius and Ghana. Transactions of the Royal Society of Tropical Medicine and Hygiene 66:108–118.

Wright CA, Ross GC. 1980. Hybrids between *Schistosoma haematobium* and *S. mattheei* and their identification by isoelectric focusing of enzymes. Transactions of the Royal Society of Tropical Medicine and Hygiene 74:326–332.

Wright CA, Ross GC. 1983. Enzyme analysis of *Schistosoma haematobium*. Bulletin of the World Health Organization 61:307–316.

Wright CA, Southgate VR. 1976. Hybridization of schistisomes and some of its implications. Pp. 55–86 In: A.E.R. Taylor and R. Muller, Eds. Genetic Aspects of Host-Parasite Relationships. 14th Symposium of the British Society of Parasitology. London. Blackwell Scientific Publications.

Wright CA, Southgate VR. 1981. Coevolution of digeneans and molluscs, with special reference to schistosomes and their intermediate hosts. Pp. 191–205 In: P.L. Forey, Ed. The Evolving Biosphere. Cambridge\ Cambridge University Press.

Wright CA, Southgate VR, Knowles RJ. 1972. What is *Schistosoma intercalatum* Fisher, 1934? Transactions of the Royal Society of the Tropical Medicine and Hygiene 66:28–56.

Wright CA, Southgate VR, van Wijk HB, Moore PJ. 1974. Hybrids between *Schistosoma haematobium* and *S. intercalatum* in Cameroon. Transactions of the Royal Society of Tropical Medicine and Hygiene 68, 413–414.

Wright MD, Tiu WV, Wood SM, Walker JC, Garcia EG, Mitchell GF. 1988. Schistosoma mansoni and Schistosoma japonicum worm numbers in 129/J mice of two types and dominance of susceptibility in Fl hybrids. Journal of Parasitology 74:618–622.

Wu GY, Wu CH, Dunn MA, Kamel R. 1985. Stimulation of *Schistosoma mansoni* oviposition in vitro by animal and human portal serum. American Journal of Tropical Medicine and Hygiene 34:750–753.

Wu LY. 1953. A study of the life history of Trichobilharzia cameroni sp. nov. (family Schistosomatidae). Canadian Journal of Zoology 31, 351–373.

Xu YZ, Dresden MH. 1989. Schistosoma mansoni: egg morphology and hatchability. Journal of Parasitology 75:481–483.

Xu YZ, Dresden MH. 1990. Minireview: The hatching of schistosome eggs. Experimental Parasitology 70:236–240.

Yamaguti S. 1958. Systema Helminthum. Volume I. Digenetic Trematodes. New York. Interscience Publishers.

Yamaguti S. 1971. Synopsis of Digenetic Trematodes of Vertebrates. Tokyo. Keigaku Publishing Company. Volume 1, 1074 pp, Volume 2, 1976. Figures on 349 plates.

Yang EP, Chow L. 1981. Acid protease in *Schistosoma japonicum* (Formosan strain). Chung Hua Min Kuo Wei Sheng Wu Chi Mien I Hsueh Tsa Chih. 14:179–186.

Yasuraoka K, Irie Y, Hata H. 1978. Conversion of schistosome cercariae to schistosomula in serum-supplemented media, and subsequent culture in vitro. Japanese Journal of Experimental Medicine 48:53–60.

Yen TC, Chiu LZ, Hsueh HC. 1974. The development and changes of the eggs of

Schistosoma japonicum in the tissues of animals. Acta Zoologica Sinica 20:272.

Yolles TK, Moore DV, Meleney HE. 1949. Post-cercarial development of *Schistosoma mansoni* in rabbit and hamster after intraperitoneal and percutaneous infection. Journal of Parasitology 35:276–294.

Yoshino TP, Bayne CJ. 1983. Mimicry of snail host antigens by miracidia and primary sporocysts of Schistosoma mansoni. Parasite Immunology 5:317–328.

Yoshino TP, Cheng TC. 1978. Snail host-like antigens associated with surface membranes of Schistosoma mansoni miracidia. Journal of Parasitology 64:752–759.

Zanotti EM, Magalhaes LA, Piedrabuena AE. 1982a. Morfologia e desenvolvimiento de *Schistosoma mansoni* Sambon, 1907 em infecções unisexuais experimentalmente produzidos no camundongo. Revista de Saude Publica de São Paulo. 16:114–119.

Zanotti EM, Magalhaes LA, Piedrabuena AE. 1982b. Localização de Schistosoma mansoni no plexo portal de Mus musculus experimentalmente infectados por um só sexo do trematódeo. Revista de Saude Publica de São Paulo 16:220–232.

Zibulewsky J, Fried B, Bacha WJ Jr. 1982. Skin surface lipids of the domestic chicken, and neutral lipid standards as stimuli for the penetration response of Austrobilharzia variglandis cercariae. Journal of Parasitology 68:905–908.

Index

Note: Table contents are not indexed. f = figure; t = table